Also by Matthew Brzezinski

· · ·

Fortress America:
On the Front Lines of Homeland Security

Casino Moscow:
A Tale of Greed and Adventure
on Capitalism's Wildest Frontier

RED MOON RISING

RED MOON RISING

SPUTNIK AND THE HIDDEN RIVALRIES THAT IGNITED THE SPACE AGE

MATTHEW BRZEZINSKI

TIMES BOOKS

HENRY HOLT AND COMPANY ● NEW YORK

Times Books
Henry Holt and Company, LLC
Publishers since 1866
175 Fifth Avenue
New York, New York 10010
www.henryholt.com

Henry Holt® is a registered trademark of
Henry Holt and Company, LLC.

Library of Congress Cataloging-in-Publication Data

Brzezinski, Matthew, 1965–
    Red moon rising : Sputnik and the hidden rivalries that ignited
the Space Age / Matthew Brzezinski. — 1st ed.
        p. cm.
    Includes bibliographical references and index.
    ISBN-13: 978-0-8050-8147-3
    ISBN-10: 0-8050-8147-X
1. Space race—United States—History—20th century.   2. Space race—
Soviet Union—History—20th century.   3. Cold War.   4. Sputnik
satellites.   5. United States—Politics and government—1953–1961.
6. Soviet Union—Politics and government—1953–1985.   I. Title.
TL789.8.U5B784   2007
629.4'109045—dc22   2007008227

Henry Holt books are available for special promotions
and premiums. For details contact: Director, Special Markets.

First Edition 2007

Designed by Victoria Hartman

Printed in the United States of America

10   9   8   7   6   5   4   3   2   1

To Lena,
who wants to be a scientist
and do "science stuff"

CONTENTS

RED MOON RISING

PROLOGUE

The rocket rose slowly, tentatively at first, as if reluctant to break from its earthly moorings. Three full seconds elapsed before its tail fins finally cleared the mobile gantry crane that had held it upright.

But now that it was free of its tethers, it seemed to lose self-doubt. Already it felt lighter, stronger. It had consumed nearly 1,000 pounds of propellant to travel those difficult first fifty feet, and the laws of physics were kicking in. Every second from now on meant 275 fewer pounds of spent fuel and oxidizer to carry. Momentum was entering the equation, increasingly on the rocket's side. Now it took only a fraction of a second to ascend the next fifty feet; even less to climb the fifty after that.

The worst was over. The heaviest lifting behind, the rocket seemed to settle down; the shaking and shuddering eased, the strain on its structure and systems diminished. It was building off its own forward progress, climbing higher and higher. Beneath it, the clearing in the forest from where it had been launched began to recede from view. Soon the firing site's fuel tankers, transport trailers, sixteen-ton Strabo crane, and armor-plated command-and-control trucks disappeared underneath the canopy of Dutch pines.

The rocket gained speed, rising at 500, 600, and then 700 feet per second. Below, lights from the nearby horse-track oval outside suburban Wassenaar emerged in the evening dusk. Moments later, the glow from the larger city of The Hague filled the panorama. Still, the rocket climbed higher over the Dutch coast, maintaining an exact ninety-degree ascent to prevent its fragile aluminum fuel bladders from bursting. To save weight, they had been designed with a thin skin that could support the rocket's 18,948 pounds of propellant only in the vertical position. Any pitch or roll at this stage and the shifting fuel would rupture the holding tanks and explode.

The rocket needed time to lose weight. And so it stayed on its upward course, its trajectory traced by a billowy white vapor trail as its 580-horsepower hydrogen-peroxide steam turbine pumped fuel through 3,184 injection ports into the combustion chamber. There, an electric igniter, a spinning wheel of spark plugs, set fire to the misty mix of ethyl alcohol and liquid oxygen, producing a vein of jet exhaust gases at 4,802 degrees Fahrenheit. That superheated exhaust—exiting through a narrow nozzle and sprayed with an alcohol propellant to increase thrust—pushed the rocket ever faster. Its rate of ascent was now 1,100 feet per second, its altitude two miles.

Far below, the 117-member firing crew of the second battery of the 485th Artillery Battalion under the command of SS Gruppenführer Hans Kammler suddenly heard a sound like the cracking of a giant whip. They knew without looking up that their rocket had just broken the sound barrier. And it was still accelerating, outrunning its own blast wave.

Twenty-five seconds had elapsed since liftoff. During that period, the rocket had shed six thousand pounds. The temperature of the sheet metal covering parts of its outer skin had spiked from 297 degrees below zero Fahrenheit at the time of fueling to over 300 degrees. The stress on its steel rib cage and interior framing had increased to nearly four times the force of gravity. But most important, its two delicate fuel bladders had drained by more than a third, allowing it to safely begin maneuvering.

At last the rocket was truly free. As if emboldened by its liberation from a strictly upward trajectory, it gained even more speed and altitude as it rolled gracefully on its axis, veered west, and bowed its long, tapered body in a forty-five-degree inclination. Ten miles below, the

Dutch coastline retreated from view, giving way to the inky blackness of the North Sea. The rocket was now traveling at twice the speed of sound, still climbing. Vanes in its four tail fins adjusted trim, pitch, and yaw, aided by heat-resistant graphite rudders that directed the flow of its white-hot exhaust gases. The steering mechanisms were in turn guided by a gyroscopic platform fitted beneath the nose cone. This inertial navigation system—the world's first—was the rocket's brain, the genius that separated it from all other pretenders. The gimbaled spinning wheels, rotating at 2,000 revolutions per minute, pointed in the same direction no matter where the rocket moved. Accelerometers attached to gimbals read the rocket's rotations relative to the rigid platform, telling it exactly where it was heading and what course to steer. But though the rocket had freedom of motion, it did not have free will. Its gyros were preset, its trajectory preordained by a complex set of mathematical equations.

Now, those calculations told it to reduce its trim, to increase its inclination to forty-nine degrees. It did so with the flick of a fin, a tiny twist of the rudder. The new bearing proved an easier angle of ascent, allowing it to go faster still. As thousands more pounds of propellant burned off, the rocket's thrust-to-weight ratio, so adverse at liftoff, grew increasingly favorable. Now that its motor was no longer pushing such a heavy load, it could really let loose, plowing forward at 3,335 miles an hour.

Fifteen miles above sea level, the rocket entered the upper layers of the atmosphere, where combustion engines normally suffocate in the thin air. But it breathed easily because it carried its own supply of liquid oxygen. The negative G forces generated by its sustained acceleration now generated eight times the pull of gravity, compressing its nose and ribs. Soon, its skin began to itch from atmospheric friction, heating in places to 500 degrees. But it felt light as a butterfly: it had dropped 17,000 of its original 27,000 pounds and was moving a mile a second.

Sixty-three seconds into its flight, the rocket ceased being a rocket. At an altitude of seventeen miles, the turbine shut down, cutting off fuel to the combustion chamber. Now the rocket was a projectile, a forty-six-foot-long artillery shell painted in a jagged camouflage scheme of signal white, earth gray, and olive green. Although no longer under power, the rocket still rose, moving at 3,500 miles per hour. The laws of physics that had gotten it this far would see it through the final leg of the journey. It broke the twenty-mile and then the thirty-mile barrier, and continued to climb. Another ten seconds passed, and the rocket reached its apogee of fifty-two miles. It was now brushing against the void of outer space.

Slowly, imperceptibly at first, its tail began to dip. It still hurtled forward at nearly five times the speed of sound, but it was now losing altitude and velocity, and its stabilizer fins had no grip as it tumbled backward toward earth. The tug-of-war between the forces of momentum and gravity had begun.

The time was 6:41 PM, a Friday. In London, fifty miles below and eighty miles to the west, the evening rush hour was just waning. Traffic was not heavy, owing to gasoline rationing, and the city was settling down to supper. Lights flickered from apartment windows and homes; the blackouts that had been in effect during the air raids were no longer rigorously enforced now that the Royal Air Force had repelled the Luftwaffe and claimed control of the skies. In central London, an American radio reporter by the name of Edward R. Murrow prepared his evening broadcast. In Chiswick, in the western suburbs, six-year-old John Clarke was freshening up for dinner in the upstairs bathroom of his family's brick town house. His three-year-old sister, Rosemary Ann, was playing in her bedroom down the hall. Both children were in good spirits. After all, there was no school tomorrow.

Over the North Sea, the rocket was now in freefall, plunging tail-first at a rate of 3,400 feet per second. As it descended into the thicker atmosphere, its fins began regaining traction and slowly the rocket righted itself. The aerodynamic friction that permitted its stabilizers to regain their function now began to wreak havoc on its rapidly heating shell. The nose cone glowed faintly as the temperature of the quarter-inch-thick sheet metal that encased it rose to 1,100 degrees. The frictional drag increased as the rocket descended into the lower, denser atmosphere and approached the British coastline. It shot over Ipswich and Southend-on-Sea, pivoting slightly, so that the jagged edges of its camouflage pattern twirled in a grayish-green blur. Other than its protective paintwork, and a seven-digit serial number, the rocket bore no identifying markings. To its creator, the brilliant engineering prodigy

Wernher von Braun, it was known as the A-4. To the Nazi High Command it was the Vergeltungswaffen-2, the new Vengeance Weapon that would restore the balance of power in the air campaign. The British would come to know it as the V-2, the world's first ballistic missile.

Now in the final moments of its four-minute flight, the V-2's target area sprawled out beneath it in every direction. The rocket was blind. Its navigation system had stopped working at the seventeenth-mile point of its journey, where it had been hurled on a predetermined trajectory. But the V-2 was an imprecise aiming device; where exactly within a ten-mile radius the missile would land was a matter of geographic chance. The East End, the City, and the Tower of London whirled past. The western part of the British capital magnified into view. The rocket's target range grew narrower. A neighborhood, Chiswick, now loomed ahead. Its red-tile roofs and cobblestone streets approached at three times the speed of sound. But no one heard the V-2 coming. Its sonic boom, that shrill thundering crack that sounded like a nearby lightning strike, hadn't caught up to it yet.

Silently, the V-2 slammed into Staveley Road at Mach 3, gouging a crater thirty feet long and eight feet deep. A millisecond after impact, its 1,627 pounds of high explosives, a pink, puffy mixture of ammonium nitrate and Amatol, detonated.

The explosion and ensuing sonic boom deafened John Clarke. His parents' house, and the homes of six of their neighbors, crumbled around him. The bedroom walls parted and fell away. Floors imploded in heaps of dust and plaster. Bricks and wood splinters crashed through windows like shrapnel. Furniture flew. Ceilings collapsed. Hallways caved in. And all this occurred in an eerily noiseless vacuum. "The best way to describe it is television with the sound off," Clarke told the BBC sixty years later.

When the mushroom cloud over Staveley Road dissipated, Clarke saw that the bedroom where his sister had been playing stood intact. Miraculously, his sister also seemed unharmed. "There wasn't a mark on Rosemary," he recalled. He shook her, but she didn't respond. The blast wave had collapsed her little lungs. She had died where she sat.

Two minutes later, another V-2 struck North London, killing six more people. That night, Edward R. Murrow informed his American

listeners of a new German weapon that rained "death from the stratosphere."

"German science," he predicted in a subsequent broadcast in November, after 168 people perished when a Woolworth department store
suffered a direct V-2 hit, "has once again demonstrated a malignant ingenuity which is not likely to be forgotten when it comes time to estab-
lish control over German scientific and industrial research."

• • •

The war was over—at least in the technology corridor of Adlershof,
just outside Berlin. Marshal Georgy Zhukov's tanks and the First
Ukrainian Red Army Group had rolled through the industrial suburb
several days earlier, on April 26, 1945, and the fighting had been brief.
Artillery still echoed from the not-too-distant German capital, where
the Führer had gone to ground in his bunker under the Reich Chan-
cellery, but in Adlershof residents were already clearing debris from
streets, filling in the bomb craters along Rudower Chausse, and carting
away glass from shattered storefronts. There was a sense of relief and
resignation throughout the town, as if its inhabitants had already made
peace with a new era and master.

Boris Chertok had no trouble finding the big brown brick building.
It was exactly where Soviet military intelligence said it would be.
Downshifting his commandeered gray Mercedes, Chertok pulled up to
the entrance. The main gate had been ripped off its hinges, and a body
lay slumped near a twisted bicycle stand. But otherwise the research
center seemed undamaged. Cautiously, Chertok made his way inside.
He was nervous. Berlin's new masters were still jumpy, unnerved by the
perplexing sight of ordinary Germans calmly tending their lawns and
rose gardens as Soviet T-34 tanks clattered past their homes to blast the
Reichstag.

Unholstering his pistol, Chertok stepped into the darkened complex.
Everything was intact: equipment, safes, precision tools, files with test
results, and all manner of documents and blueprints, many stamped
SECRET or TOP SECRET. Even the keys to the gleaming white laboratories
had been left behind—obligingly numbered, in neat orderly rows. Chertok stowed the gun. He barely knew how to use it anyway. The sidearm,
like his ill-fitting uniform, was a new and uncomfortable acquisition, a

recent addition to his wardrobe that left him feeling like an imposter. The bars and gold star on his shoulder boards identified him as a major in the Red Army, but the military garb would not have fooled a seasoned veteran. The tunic was too clean, the boots too unscuffed for a frontline officer. What's more, not a single combat decoration hung from Chertok's breast, a highly suspect omission after four years of all-out war. Major Chertok, it was clear, was only masquerading as a soldier. In reality he was a thirty-three-year-old electrical engineer with a prematurely receding hairline, the slight paunch of someone who spends too much time behind a desk, and a dossier identifying him as an employee of NII-1, the Soviet Union's leading rocket research agency. Hundreds of civilian specialists just like him were pouring into Germany, arriving daily in a motley assortment of Dodges, Studebakers, and converted Boston B-25 bombers that the United States had given the Soviet Union under the lend-lease program. And they were coming—as Edward R. Murrow had prophesied—to claim the intellectual spoils of war, to seize upon Nazi Germany's "malignant ingenuity."

The research institute in Adlershof was just one in a burgeoning catalog of places of scientific interest in the Berlin area. There was the Askania factory nearby, the Siemens plant in Spandau, a design bureau in Mariendorf, another Askania facility in Friedenau, a Telefunken factory in Zehlendorf, and the list went on. Each of these sites held its own wonders: magnetron tubes with a pulse power of up to one hundred kilowatts, accelerometer calibrators, polarized relays, transverse and longitudinal acceleration integrators—precision instruments that the Russians had only dreamed of. It was such sophisticated components that made the world of difference between the giant, guided V-2 and the crude little directionless Katyushkas the Soviet Union had built during the war. And Joseph Stalin wanted that technological edge.

German precision engineering was now the official property of the Soviet government. The specialists back home were anxiously awaiting the war trophies. Epic bureaucratic battles would later erupt in Moscow over who got the booty. Fortunately for Chertok, his initial role in this national technology transfer was restricted to inventorying, packing, and shipping the loot. Others would have to fight over it. At times he felt like a wide-eyed child that had been handed the keys to the Detsky Mir toy store across from NKVD headquarters in Dzerzhinski Square.

"Oh, this German love for details and this exactness, which has in-grained such top-notch work into the culture," he declared in an April 29 diary entry. "I am envious."

Anything was for the taking—exquisite Khulman drafting tables, vibration benches, entire photochemical laboratories. And what brand names: Philips, Rohde & Schwartz, Lorentz, Hartmann-Braun, Haskle, AEG, Karl Zeiss. Chertok had read about these famous firms in the Western journals that were selectively circulated at NII-1, but until now they had been as familiar and unavailable as a Lana Turner pinup.

Taking it all seemed only fair. "We have every right to this," Armaments Minister Dmitri Ustinov explained. "We paid for it with a great deal of blood." After all, it was the Germans who had violated the 1939 Molotov-Ribbentrop nonaggression pact, who had attacked unprovoked two years later and slaughtered millions of Soviet citizens. In Russia, they lacked everything, while Germany had an overabundance of everything. "The thing that every laboratory needs the most and is in the shortest supply," Chertok continued in his diary, "is the Siemens four mirror oscillograph. In Moscow, at NII-1, we only had one for the entire institute. And these Germans had so many!"

Surveying the marvels of German science, Chertok could not help being swept up in the moment, pushing aside the fact that the Nazis had murdered his Jewish relatives in Auschwitz and ravaged his homeland. "No," he wrote on April 30, "we no longer felt the hatred or thirst for revenge that had boiled in each of us earlier. Now it was even a pity to break open these high-quality steel laboratory doors and to entrust these diligent but not very careful soldiers with packing priceless precision instruments. But faster, faster—all of Berlin is waiting for us! I am stepping over the body of a young *panzerfaust* operator that has not yet been cleared away. I am on my way to open the next safe."

• • •

Five hundred miles west of Berlin, ensconced in the Parisian luxury of the Plaza Athénée Hotel, Colonel Holger N. Toftoy had been issued almost the same instructions as Boris Chertok. Only his scavenging list consisted of one item: the V-2.

Like Moscow, Washington wanted the rocket. The U.S. Army had quickly grasped that the V-2 represented a new type of weapon that could revolutionize warfare. The rocket had demonstrated that it could deliver significant explosive charges over long distances with relative accuracy. It could be mass-produced (Hitler had ordered twelve thousand units made) and easily transported. Its potential was obvious. It was even possible that guided missiles might someday replace artillery and make long-range bombers obsolete. The U.S. brass was not about to pass up the chance to grab this promising new weapon, as Toftoy's boss, Major General Hugh Knerr, made clear in a 1945 memo. "Occupation of German scientific and industrial establishments has revealed the fact that we have been alarmingly backward in many fields of research," Knerr wrote. "If we do not take this opportunity to seize the apparatus and the brains that developed it . . . we will remain years behind. Pride and face-saving have no place in national insurance." Knerr's boss, General Dwight D. Eisenhower, put it even more starkly in a cable to Washington. "The thinking of the scientific directors of this group is 25 years ahead of U.S. Recommend that 100 of the very best men of this [V-2] organization be evacuated to U.S immediately."

Toftoy's marching orders were equally terse. "Get enough V-2 components to make 100 complete rounds," he instructed in the spare tone preferred by the military. "Ship to US." Attached to the directive was a "black list" of names headed by Wernher von Braun, the boy wonder who at the age of twenty-four had been put in charge of what would become the Third Reich's most important military project. It sounded so simple: round up one hundred rockets and one hundred men. Little did Toftoy know that the assignment would preoccupy the next ten years of his life.

Unlike Chertok, "Ludy" Toftoy (as he was oddly nicknamed by his classmates at West Point) was a career military officer, a graduate of numerous advanced programs at the Command and General Staff School, the Ordnance School, and the Army-Navy Staff College. He was a crack marksman, a recipient of the Knox artillery trophy and the Distinguished Pistol Shot medal. Slim and square-jawed, Toftoy was the embodiment of the American soldiering spirit. He had brains and brawn, and an innate talent for organization, which he displayed amply and eagerly from his earliest days at West Point, where he was chairman of the Ring Committee, designer of the 1926 Class Crest, president of the Dialectical Society, art editor of the student newspaper, and a four-year letterman on the pistol team. The West Point yearbook, the *Howitzer*, summed up his busy stay on the banks of the Hudson: "It is no exaggeration to say that almost everything that [the class of] '26 has done during the past two years has felt the guiding hand of little Ludy."

Little Ludy, now grown up and newly promoted to chief of Ordnance Technical Intelligence in the European theater, was for once having difficulty carrying out his orders. Though the V-2s he was supposed to find had been mostly fired from mobile launch sites in Holland and France, there were no rockets in either country. The missiles, apparently, had not been designed for lengthy storage periods, and firing batteries like the 2./485 that had launched the first strike against London tended to run through the German rocket supply faster than factories could replenish it. So Toftoy had to go to the source, which was problematic since the First Belorussian Red Army Group had overrun the V-2's original production facility, a sandy islet in the Baltic Sea called Peenemünde. Fortunately, Peenemünde was of limited value to the Soviets because the SS had moved the V-2 assembly lines to a secret location after a series of Allied air raids in August 1943.

Despite Toftoy's numerous and increasingly frantic entreaties, army intelligence still hadn't located the new facility. It could be anywhere in Germany or Austria. Worse, it could be in Soviet-occupied territory, like Peenemünde, in which case Toftoy would never be able to carry out his mission. As Ludy contemplated the unpleasant prospect of a strikeout blemishing his stellar service record, a message clattered off the Teletype machine at G-2 army intelligence headquarters in Frankfurt. The 104th "Timberwolf" Infantry Division had apparently stumbled across something horrible in the Harz Mountains in central Germany.

John M. Galione, a private with the 145th Regiment, had made the grisly discovery by literally following his nose. "Hey Sarge, what do you think that odor could be?" he asked early on the morning of April 5. The smell was the stench of decaying corpses from a passing prisoner transport train. For four days, Galione followed the rail tracks, convinced they led to a concentration camp. He walked more than one hundred miles, mostly at night and by himself. "Something overpowering came over me," he recalled fifty years later. "I don't know what it was. My legs just kept walking. It was as if someone was pushing me from behind."

Finally, early on April 10, after hiking again all night, he came across an abandoned railcar shortly before dawn. It was filled with dead bodies. Behind it, some sort of a camp emerged in the morning mist. As the sun rose, piles of dead bodies grotesquely materialized. "They were gray in color, and they looked like skeletons wrapped in skin. Some of them were so thin you could see their backbones through their stomachs."

But amid the horror, something else caught Galione's eye. "From where I was standing, I could see a hidden tunnel coming out of the side of the mountain. That's how I knew I had found something big the Nazis were trying to keep secret."

What Galione had found was Mittelwerk, the giant subterranean V-2 factory, and Dora, its attendant concentration camp.

When word of the find reached Toftoy at the Plaza Athénée Hotel a few days later, he immediately grabbed a map. Mittelwerk was near Nordhausen, in Thuringia. *Oh my God*, he thought. That region of Germany was slated to become part of the Soviet occupation zone. His closest inspection team was in Fulda, eighty miles away. It would take weeks just to inventory the more than one million square feet of caverns that connected Mittelwerk's two-mile-long assembly-line tunnels. Subassemblies with tens of thousands of often highly complex component parts for fuel pumps, guidance systems, electronic relays, and everything else that made a rocket fly would need to be carefully examined and cherry-picked. The few intact rockets that had been completed just before the Nazi evacuation would have to be partly dismantled and readied for transport. Specialized troops with the necessary mechanical skills to handle such fragile cargo would need to be found. The only unit that remotely fit the bill was the 144th Motor Vehicle Assembly Company. And it was currently in Cherbourg, France, 770 miles away. And, as if all that wasn't challenging enough, German engineers that could put all this stuff back together in working order would need to be located and their cooperation enlisted. Months could pass for an operation of this scale to be mounted.

Don't touch anything, Toftoy ordered. I'm on my way.

. . .

When Chertok and the Russians arrived at Mittelwerk, on July 14, 1945, the place had been virtually picked clean. The Americans had hauled away one hundred intact rockets and had filled sixteen Liberty Ships with 360 metric tons of component parts for transfer to the White Sands Proving Ground in New Mexico. They had made off with an entire liquid oxygen plant, all sorts of fueling equipment, static test-firing rigs, and a dozen mobile launchers. Most of Mittelwerk's precision tools and sensitive bench instrumentation panels were gone, as were the German engineers who knew how to operate them. Even the giant overhead tunnel lights that illuminated the underground production lines had apparently been tinkered with, so that Chertok and his team were left to grope dangerously in the dark.

"The problem is this," another Soviet rocket scientist, Colonel Grigori A. Tokady, glumly reported to Moscow. "We have no leading V-2 experts in our zone; we have no complete projects or materials of the V-2; we have captured no fully operational V-2s which could be test launched right away. We [only] have lots of bits and pieces."

At least the Americans had buried the dead at Dora—nearly five hundred Russians, Poles, and Hungarian Jews—and nurtured the skeletal survivors back to a semblance of life. Still, Chertok could barely bring himself to visit the concentration camp, where mounds of human ash still littered the crematorium, and the horrors of working at Mittelwerk still haunted the hollow features of its freed slave laborers. The majority of the factory's twenty thousand victims had been Russian POWs—often hanged a dozen at a time for minor infractions from the overhead crane that ran the length of Tunnel B. A French prisoner, Yves Beon, described being forced to file past the corpses dangling above the V-2 assembly line: "Most of their bodies have lost both trousers and shoes, and puddles of urine cover the floor. Since the ropes are long, the bodies swing gently about five feet above the floor, and you have to push them aside as you advance. . . . You receive bumps from knees and tibia soaked in urine, and the corpses, pushed against each other, begin to spin around. . . . Here and there under the rolling bridge, truncheons in hand, the S.S. watch the changing of the shifts. They are laughing: it is a big joke to these bastards."

For each of the 5,789 V-2s produced at Mittelwerk, which had been built by the same engineer who designed Auschwitz, nearly four prisoners had died. But those who survived now came forth to offer their intimate knowledge of the factory's intricate cavern network. One former prisoner told Chertok, "I know places where the SS hid the most secret V-2 equipment that the Americans didn't find." As Chertok later recalled, "He led us to a distant wooden barracks hut, where in a dark corner, after throwing aside a pile of rags, he jubilantly revealed a large spherical object wrapped in blankets. I was stupefied. It was a [next generation] gyro-stabilized platform, which still hadn't become a standard V-2 instrument."

The find brightened morale among the despondent Soviet scientists. Perhaps the Americans hadn't taken everything of value after all. Toftoy had left enough exterior rocket parts—tail fins, middle-section casings, and nose cones—to piece together fifteen to twenty whole V-2 bodies. Without the innards, though, the shells were largely useless. Yet the gyroscope gave Chertok hope: if an advanced version of the V-2 guidance system could be so easily located, what other buried treasures might the Americans have missed?

Over the next eighteen months, Chertok and an increasingly crowded field of German and Soviet experts began fitting together the missing pieces of the V-2 puzzle. In nearby Bleichrode, at the bottom of a dead-end drift in a potassium mine, they found more guidance system components: Viktoria-Honnef range-control and lateral radio correction sets. In a forester's cabin, in a hunting reserve just over the demarcation line of the American sector, two sets of relay boxes and firing control panels were unearthed. One of the biggest windfalls was dug out of a sand quarry in Lehesten: fifty brand-new combustion chambers and crates with enough fuel pumps, injectors, and engine parts to fill fifty-eight railroad cars.

When the Russians first fire-tested one of those engines on September 7, 1945, the results stretched the limits of Soviet scientific imagination. The V-2 generated twenty-seven tons of thrust. The biggest Soviet rockets, by comparison, could not even manage a ton of lift. The difference amounted to a quantum technological leap. Every detail of German engine design seemed to reveal a minor engineering miracle. The Germans used alcohol as propellant, instead of kerosene mixed with liquid oxygen. The alcohol ingeniously did double-duty as a coolant for the combustion chamber to prevent meltdowns, snaking around its outer walls in coils, like a refrigeration unit. At the bottom end of the chamber the exhaust nozzles had been counterintuitively shortened and flared, creating a larger opening for the hot jet gases to exit. Such a configuration had long been rejected by Soviet designers on the theoretical principle that a larger outlet would dissipate the intensity of the escaping gases and thus reduce power. In fact, the simple and yet radical alteration increased output by 20 percent. "Pure genius," whistled Soviet propulsion specialists in evident admiration.

With several combustion chambers fully assembled, the Russians had solved part of the reverse-engineering puzzle. But twenty thousand separate parts went into each V-2, and correctly putting the remaining pieces together into an operational guided missile was another matter. Even the scattered component parts initially had to be disassembled "down to the last screw," Chertok recalled, so that detailed drawings could be made to glean a basic understanding of how they worked.

Maddeningly, for the Russians, more than one thousand qualified German engineers were just beyond reach in the American zone, across the Werra River. To make matters worse, the engineers had hidden all the assembly instructions—six truckloads of assorted owner's manuals, including more than sixty thousand blueprint modification drawings—at the bottom of a mine, to be used as a bargaining chip with Toftoy, their new American benefactor. "These documents were of inestimable value," recalled Dieter Huzel, one of the German scientists entrusted with concealing the trove. "Whoever inherited them would be able to start in rocketry at the point where we had left off, with the benefit of not only our accomplishments but of our mistakes as well."

The B-team that Chertok was left with had neither the benefit of experience nor documentary guidance. It might have been numerically superior—by 1946 some six thousand Germans were on the Soviet payroll, resurrecting the V-2 program—but it consisted mostly of technicians, draftsmen, and lower-grade engineers whose knowledge rarely exceeded subassembly levels. What was missing from the Soviet effort were German scientists with knowledge of the big picture. And no amount of extra egg rations, tax-free salaries, bonuses, and bribes could entice stars of von Braun's caliber from the American side. The one exception was Helmut Grottrup, von Braun's deputy for electric systems and guidance control. Unfortunately for the Russians, he came in a package deal with the imperious Frau Grottrup, who viewed the Red

Army as her personal catering service, right down to the show horses she demanded for the stables of the sprawling villa she had selected for her residence. Frau Grottrup fired cooks and assistants on a weekly basis, and the never-ending shopping list she presented to her husband's astonished Soviet employers required the full-time attendance of a colonel. "My sister goes to university wearing men's boots," wrote one of Frau Grottrup's indignant Russian minders. "She is selling her last dress to buy food for our sick mother. My young wife Tamara had to quit her studies because she can't make it without my help—and here we are getting saddle horses!"

But Frau Grottrup got her horses, her BMW, a pair of cows for fresh milk, even her insistence on a Soviet officer as a riding companion. She got whatever she wanted because back in Moscow Stalin was furious that no other senior German scientists had come over to the Soviet side, and he was demanding that something be done. "We'd even hatched a plan to kidnap von Braun," Chertok reminisced, but the abduction efforts proved amateurish, as one U.S. intelligence report recorded: "One day, a group of men in American Army uniforms entered the schoolhouse in Witzenhausen. They began a friendly conversation with several members of the team and suggested they all go into the village for a few drinks. However the Germans were suspicious of the English they spoke—it was neither American English nor British English. The Russians left without captives."

What the Russians lacked in subtlety and senior scientists, they compensated for with an excess of Red Army zeal for pillage. As the Soviet-controlled part of Germany was systematically dismantled and shipped east, the crucial missing blueprints began turning up in the unlikeliest places: behind woodpiles, in toolsheds, in the homes of former factory directors, and in partially destroyed attics. Among the trove of documents, one set of plans particularly caught the Soviets' attention. It was a proposal for the design of a far more powerful version of the V-2, the one-hundred-ton-thrust A-10, which was to have gone into development once Hitler's European adversaries had been vanquished. The A-10 was a long-range two-stage rocket that had one purpose: to strike New York and Washington.

In Moscow, where the notion of a new war, a cold one, was already taking shape, Stalin himself quickly seized upon the idea of the A-10. "It would be an effective straitjacket for that noisy shopkeeper, Harry Truman," he advised his top officials. "We must go ahead with it, comrades. The problem of the creation of transatlantic rockets is of extreme importance to us."

1

THE REQUEST

One after another, the big ZIS limousines pulled away from the curb. Black and burly, their whitewall tires not yet soiled by slush, the armored behemoths glided gently through the snow—three-and-a-half-ton dancers in a synchronized automotive ballet.

From Central Committee headquarters on Staraya Square, the chrome-fendered procession headed east past Gorki Street and slipped under the shadow of one of the Gothic skyscrapers that Stalin had ordered built after the war. Seven of the sinister high-rises dominated the Moscow skyline, and they rose in stone layers like fifty-story wedding cakes decked in dark granite icing. On the Ring Road, another Stalinist creation, an eight-lane motorway that surrounded the city center in an asphalt moat, the dozen limousines and chase cars turned north, breezing past block after block of leviathan government buildings that heralded the new Moscow.

The old mercantile city had been branded with Stalin's indelible stamp; it was now a metropolis of bronze bas-reliefs of giant steelworkers with bulging forearms, of cast-iron tributes to collective farmers brandishing sixteen-foot scythes, of statues of Lenin six stories high. Monumental and monochromatic public works had risen on the trampled foundations of prerevolutionary pastel palaces. The gilded cathedrals of old were gone—ripped down or buried under layers of soot—and in their stead rose hulking cement fortresses whose towering porticos paid architectural tribute to the insignificance of the individual in the one-party state. Red banners hailing the Twentieth Party Congress draped the gray edifices, though the plenum had ended tumultuously two days before. GLORY TO THE WORKERS, they proclaimed. PEACE TO THE PROLE-TARIAT. Beneath the propaganda placards, long lines of shoppers formed outside food stores, their breath steaming in the cold, while elderly women in fingerless gloves swept dirty snow from the sidewalks.

If Nikita Khrushchev noticed any incongruity between the lofty slogans and the harsh socialist reality of Moscow street-corner life, he said nothing to his driver, or to his son, Sergei, sitting beside him in the backseat. He had said more than enough at the Twentieth Congress, and although six weeks would pass before the Central Intelligence Agency got wind of his secret speech, the Americans would be just as stunned as the 1,500 shocked delegates who had heard him in the Great Kremlin Palace less than forty-eight hours earlier.

Father and son sat in silence, behind the ZIS's pleated curtains, watching the city stream drably past. Directly behind them, in the lane reserved exclusively for party high-ups, rode the three other Presidium members who had jointly ruled Russia since Stalin's death three years earlier: Nikolai Bulganin, Lazar Kaganovich, and Vyacheslav Molotov. Behind them, in descending order of importance, rode lesser officials in less imposing ZIMs and Pobedas: Dmitri Ustinov, the armaments minister; Aleksei Kirichenko, a new addition to the Presidium; and the usual entourage of KGB bodyguards and eager aides.

Khrushchev's car led the convoy, as befitting his new rank as first among equals in the Presidium, as the Politburo was then known. But the men in the rear were crowding his bumper, biding their time for an opportunity to overtake him. This Khrushchev knew, understood almost instinctively, from the survival skills he had honed during two treacherous decades in Stalin's inner circle. When the "boss" was alive, they had all vied for his favor, slid tomatoes onto each other's chairs in prank-filled drinking contests that lasted till dawn. Like sixty-year-old frat boys, they all laughed when Lavrenty Beria, the secret police chief and serial rapist, pinned little notes with the Russian word *prick* spelled in big Cyrillic letters on Khrushchev's unsuspecting back. But the day the boss died, the jokes ended and the plotting began. This Khrushchev also knew from personal experience, for it was he who had master-minded the coup against the psychopath Beria. Stalin's chief henchman and self-anointed successor had been dispatched with a bullet to the brain, and it was Khrushchev, the class clown of this murderous frater-nity, who had unexpectedly emerged to fill the power vacuum.

The convoy sped north unimpeded by traffic, since so few Mus-covites owned cars. A few rickety Moskviches, twenty-six-horsepower knockoffs of the 1938 Opel Kadett, quickly clattered out of the way, and soon Khrushchev and his retinue reached the new suburbs. These vast tracts of land around the capital were also being completely re-made. The little wooden dachas and garden plots that supplemented the diet of city dwellers were being bulldozed in accordance with the latest Five Year Plan to allay the capital's dire housing shortage. In Moscow, divorced couples still often lived together in communal apart-ments with three other families. Government workers and newlyweds were housed in dormitories where toilets and kitchens were divvied up on a time-share basis. For those without *blat*—connections—the wait-ing list for new apartments could stretch into the decades. People got married while still in their teens just to get in line. The situation was much the same throughout the Soviet Union, and even more drastic in the western parts of the country, where entire cities had been reduced to rubble during the war.

To remedy the housing shortage, Khrushchev had embarked on a massive national construction binge, which, like Stalin's oppressive overhaul of Moscow's downtown districts, was leaving its own unique architectural imprint on the city's outer rings. But the suburban build-ing boom looked nothing like the Levittown planned communities sweeping across America, where the appetite for new homes seemed just as insatiable and the explosion of affordable automobiles was changing the way people lived. Outside Moscow, there were no ranch houses and ramblers with circular driveways branching off cul-de-sacs. There were no swimming pools in gated backyards; no children's play-grounds with baseball diamonds and swings. Instead, row after row af-ter row of mind-numbingly identical five-story apartment buildings rose from the sandy soil, resembling the ill-conceived housing projects that would later blight America's ghettos. Large numbers—1 through 50, 51 through 75, and so on—were painted on the sides of the squat concrete structures, so that residents could tell them apart. Not a blade of grass grew in between the prefabricated buildings of this suburban wasteland, and there were no stores, shopping centers, or movie theaters. There were only bus stops that led to the nearest commuter train station and cranes leaning over the open pits of unfinished units.

Quantity, not quality, was the construction crews' rallying cry, and the "Khrushchevki," or Little Khrushchevs, as the dwarfish blocs quickly became known, howled when the winds blew in from the Urals. They suffered from shifting foundations and perennially creaky plumbing. Their gas lines exploded and their roofs leaked. But they met Kremlin quotas, and Khrushchev was immensely proud of them. Just the other month, he had taken Sergei on a tour of the cement factory that made their prefabricated parts.

On this day, Sergei was also accompanying his father on an inspection tour. The favorite of Khrushchev's four children, Sergei, at twenty-one, already stood a head taller than his dad, whom a French official once described as a "little man with fat paws." (Khrushchev himself poked fun at his diminutive stature, joking that he was as wide as he was tall.) Sergei had his mother's looks, blond hair and blue eyes. Nikita had darker, vaguely Asiatic features, though what little hair remained on the elder Khrushchev's head had long ago turned white, leaving only a narrow band to warm his bald pate. Unlike his father, whose loquacious, self-deprecating sense of humor made him a favorite at Stalin's court (and lulled his enemies into a false sense of superiority), Sergei was a serious, studious lad, on his way to fulfilling his father's dream of becoming an engineer. The elder Khrushchev had had only four years of formal schooling before beginning his apprenticeship, at the age of fourteen, as a metalworker in a prerevolutionary Ukrainian coal mine owned by a Welsh millionaire. "After a year or two of school, I had learnt how to count to thirty and my father decided that was enough," Nikita Khrushchev recalled in his memoirs. "He said that all I needed was to be able to count money, and I would never have more than thirty rubles to count anyway."

Khrushchev's lack of education was a sore point, a source of embarrassment and frustration—not only to him but to the party as well. The leaders of the revolution had been learned men: Trotsky, Bukharin, the lawyerly Lenin. Even Stalin had studied in a seminary before finding Marx. But in the Soviet version of upward mobility, the next generation of Communist Party functionaries had risen from the bottom of the proletariat, sons and daughters of the peasantry suddenly catapulted into the twentieth century, as Khrushchev himself conceded with remarkable candor. "We weren't gentlemen in the old-fashioned sense," he wrote in his memoir, recalling his wartime stay at the estate of a Polish nobleman. "It became impossible to enter the bathroom. Why? Because the people in our group didn't know how to use it properly. Instead of sitting on the toilet seat so that people could use it after them, they perched like eagles on top of the toilet and mucked up the place terribly. And after we put the bathroom out of commission, we set to work on the park grounds."

Throughout his late twenties and thirties, Khrushchev had struggled to better himself, attending the Rabfak high school equivalency programs offered to rising party functionaries and enrolling in special courses at the Stalin Industrial Academy for promising technocrats. But party business always interfered, and he never managed to finish any of them. "He could barely hold a pencil in his calloused hand," one of his teachers later told the biographer William Taubman. "She recalled his struggling to grasp a point of grammar and, when he at last understood it, smiling and shouting, 'I got it.' "

Sergei was thus on the cusp of fulfilling his father's unrealized dream. In a few months he would defend his master's thesis and become a full-fledged engineer. "My father felt this was the best, most honorable profession a man could have," he recalled fifty years later. "In technical matters, he was very creative and curious."

Khrushchev's passion for technology could at times lead to childlike bursts of enthusiasm, and whenever state business took him to a plant or research facility of technical interest, he brought Sergei. "He wanted me to see the theories I had been learning at university applied in practice," Sergei recalled. The two had recently gone to the Tupolev factory to inspect the first Soviet jet-engine passenger plane, and Nikita had boyishly rubbed his hands in glee at the prospect of impressing foreigners with it on his next trip abroad.

Today, though, was a special outing for Sergei. Ever since the ZIS had picked him up outside class at the Moscow Institute of Power Engineering that morning, he had been giddy with anticipation. "You see, I was studying to become a rocket scientist, a guidance systems expert to be precise," he noted. And today his father was taking him to NII-88, the USSR's top-secret rocket research facility.

• • •

The design bureaus of NII-88 were discreetly tucked away outside Moscow, where too many foreigners with prying eyes roamed the streets. To get there, then as today, visitors took the main road to Yaroslavl. Khrushchev's motorcade, with the other Presidium members in tow, turned onto what is now called the M8 highway, and the cranes and suburban construction sites soon gave way to the countryside. The transition came abruptly, like crossing some imaginary threshold between the twentieth and eighteenth centuries. Roads turned to mud, settlements into ramshackle villages. Wooden farmhouses and huts with thatched roofs leaned at crazy angles. Most had no electricity or running water. Their inhabitants had few teeth. They walked around half dazed, as if in slow motion, swaddled in rags, filthy peacoats, and sleeveless jackets made from the hides of farm animals. The herds of cattle were scrawny and clumped with manure. Skinny chickens scampered underfoot.

Though the Communist Party viewed the backward peasants with undisguised contempt for both ideological and practical reasons (stemming from perennially poor harvests), Khrushchev had always felt comfortable in the countryside. He had made agriculture his bailiwick under Stalin, and he had grown up in similar circumstances, tending sheep as a young boy in the tiny farming community of Kalinovka, near Kursk. "Every villager dreamed of owning a pair of boots," he recalled. "We children were lucky if we had a decent pair of shoes. We wiped our noses on our sleeves and kept our trousers up with a piece of string."

But the massive farms that the Presidium held in such low regard played another critical role besides putting food on Soviet tables. The endless expanses they covered provided Russia with its main line of defense. It was these snow-covered fields, stretching thousands of miles, that had defeated Hitler and Napoleon. Like frozen deserts that thawed into impassable bogs, they had protected Moscow from all its

Western enemies. Armies could advance over the plains, but invariably their supply lines would grow thin, the winter would set in, and rural Russia would ravage the invaders. The steppe had always afforded Moscow the ultimate victory. Until now. Now, in the nuclear and jet age, distance and climate no longer provided a natural limit to foreign depredations. And to Khrushchev, that seemed precisely what the latest U.S. military doctrine aimed to do.

Khrushchev was unsettled by the rise to power of the Republican Party, after more than two decades of Democratic rule. The Republicans represented the American capitalist class, and their electoral battle cry had been hard-line anticommunism. To the Soviets, the emergence of the rich, Russophobe Republicans signaled the arrival of a more combative and ideological adversary in Washington, personified by John Foster Dulles, the dour and deeply religious secretary of state, a man who dressed and talked like a clergyman and yet managed to make millions during the Great Depression. The USSR, Dulles declared, could never be appeased, because "the Soviets sought not a place in the sun, but the sun itself." His opinion was codified by the National Intelligence Estimate of September 15, 1954, which stated, "Soviet leaders probably envision: (a) the elimination of every world power center capable of competing with the USSR; (b) the spread of communism to all parts of the world; and (c) Soviet domination over all other communist regimes."

With growing alarm, the Soviets watched as Dulles purged the State Department of suspected liberals. Veteran foreign service officers who had accurately predicted Communist gains in Asia were sacked for not displaying "positive loyalty," while foreign allies were warned that they too had better toe the new hard line or be faced with an "agonizing reappraisal" of U.S. assistance. Dulles's playboy brother Allen, whose hedonism was matched only by his hatred of communism, was put in charge of the Central Intelligence Agency, which rapidly ballooned from an obscure bureaucratic outpost with 350 employees to an aggressive frontline agency with thousands of operatives intent on undermining Soviet power.

John Foster Dulles lurked dangerously behind the kind, grandfatherly facade of President Dwight D. Eisenhower, whom the Soviets knew to be ill with a heart condition and increasingly detached from day-to-day affairs of state. It was the unelected and standoffish lawyer, not the popular war hero, who thus dictated U.S. policy. America's moral duty, Dulles declared, was not merely to stop the spread of communism but to "liberate captive peoples" all over the world. The Eisenhower administration, Dulles pledged, would "roll back" Communist advances in Europe and Asia and send the Soviets packing. What's more, he continued, the United States would no longer bother with small local conflicts like Korea to keep communism in check. Henceforth it would prepare for "total war," a phrase coined by Admiral Arthur Radford, the chairman of the Joint Chiefs of Staff, and wage an "instant, tremendous, and devastating" nuclear attack on the Soviet Union itself. Only a doctrine of "massive retaliation" promised "to create sufficient fear in the enemy to deter aggression." The strategy, Dulles noted, "will depend primarily on a great capacity to retaliate instantly by means and at places of our choosing."

To the stunned Soviets, who did not yet have the effective capacity to launch any sort of surprise attack on the United States (as Dulles well knew), the massive retaliation doctrine was perceived as little more than a massive intimidation tactic. "We shall never be the aggressor," Eisenhower had reassured the Russians at a summit meeting in Geneva in 1955, but Khrushchev had no guarantees of that. The only way for Khrushchev to guarantee Soviet security was to develop his own massive retaliation capabilities. But he lagged far behind his American rivals.

Nuclear weapons production in America had been ramped up to an industrial, assembly-line scale under the Eisenhower administration. By 1955 the United States had amassed 2,280 atomic and thermonuclear bombs, a tenfold increase from 1951, representing an arsenal nearly twenty times greater than the Soviet stockpile. (As Dulles's doctrine evolved, the number of warheads would jump to 3,500 by late 1957, double to 7,000 by 1959, hit 12,305 by 1961, and top 23,000 two years later.) Meanwhile, billions of dollars were being poured into an armada of heavy long-range bombers to deliver the nuclear payloads. By 1956 the air force bomber fleet had almost doubled in size, and the Strategic Air Command kept a third of its 1,200 B-47 long-range bombers on the runway at all times, fueled and loaded with their nuclear cargo. Curtis LeMay, the cigar-chomping SAC commander, seemed to be on a personal mission to instill fear in Russian hearts. In January 1956, LeMay scrambled almost all his bombers in a simulated nuclear attack. In another exercise, Operation Powerhouse, his planes flew nearly one thousand simultaneous sorties from more than thirty bases around the world to intimidate Moscow. In a few weeks, he would launch yet another exercise called Operation Home Run—reconnaissance versions of his B-47 Stratojets would fly from Thule, Greenland, over the North Pole, and into Siberia to probe for gaps in Soviet radar defenses. The mission would culminate with a squadron of the metallic silver RB-47s, their undersides painted white to reflect the flash of a nuclear blast, flying in attack formation in broad daylight several hundred miles into Soviet territory. The Soviets would have no way of knowing that the bombers were not armed, or that an attack was not imminent. And that would be the point of the exercise: to expose the USSR's defenselessness against a polar attack and to drive home the message that the United States could strike Russia at will. "With a bit of luck, we could have started World War III," LeMay would later reminisce ruefully.

At times LeMay's antics even scared the CIA. "Soviet leaders may have become convinced that the US actually has intentions of military aggression in the near future," warned an ad hoc committee of CIA, State Department, and military intelligence agency representatives. "Recent events may have somewhat strengthened Soviet conviction in this respect."

From their American bases in Greenland, Norway, Germany, Turkey, Britain, Italy, Morocco, Pakistan, Korea, Japan, and Alaska, B-47s could reach just about any target in the Soviet Union, furthering LeMay's well-publicized goal of obliterating 118 of the 134 largest population and industrial centers in the USSR. (LeMay calculated that 77 million casualties could be expected, including 60 million dead.) And he was about to get an even bigger bomber, the intercontinental B-52 Stratofortress, which was just entering into service. The giant plane could carry 70,000 pounds of thermonuclear ordnance over a distance of 8,800 miles at a speed of more than 500 miles an hour. With the B-52, the Americans no longer even needed their staging bases in Europe and Asia to attack Russia. They could do it from the comfort of home without missing more than a meal.

Most distressing for Khrushchev, he had no way of striking back. The biggest Soviet bomber in service, the Tupolev Tu4, was an aging knockoff of the propeller-driven Boeing B-29 with a 2,900-mile range and no midair refueling capacity. It could not effectively reach U.S. soil. The Tu4 would either run out of gas as it approached the American eastern seaboard or crash in the coastal states of New England. In either scenario, planes and pilots would be lost on one-way suicide missions. Unfortunately for the Kremlin, the early prototypes for a pair of bigger bombers, the Mya-4 Bison and Tu95 Bear, which were designed to hit targets deep in U.S. territory, seemed to display similarly suicidal tendencies. Their test flights had been plagued by crashes, and it would be years before they were operational in significant numbers.

The bottom line was that the United States could stage a multipronged attack on the USSR from dozens of points across the globe, while the Soviet Union was hemmed in from all sides and could not retaliate. It was this strategic imbalance, and the urgent need to redress it with an effective retaliatory capability of their own, that drove Khrushchev and the other Presidium members through the windswept countryside on February 27, 1956, to visit the secret missile laboratories of NII-88.

• • •

The tiny czarist town of Podlipki had been erased from Soviet maps. With a swipe of the pen and an eye toward subterfuge, Kremlin cartographers had rechristened it Kaliningrad, the same name they had given to a large Baltic enclave seized from Germany after the war. Such attempts at misdirection were common to conceal sensitive installations, aimed at sending American spies a thousand miles the wrong way, but this one didn't fool the CIA for long. By the mid-1950s, German V-2 scientists repatriated from Russia had already given the agency a vague idea of what was going on behind the birch forests and high fences that hid NII-88. Allen Dulles knew, for instance, that to get there visitors had to take a series of right turns along a maze of unmarked country roads; that the approach was discreetly monitored; and that about half a mile down the main perimeter wall, a narrow automated steel gate known as the "Mousetrap" governed access to the grounds.

It was through the Mousetrap that Khrushchev's ZIS-110 slid. In a few months, he would have a new limo—the sleeker, squarer ZIL, modeled after the 1954 Cadillac—and the first party secretary was anxiously awaiting its delivery. In addition to the modern redesign, the ZIL held the promise of being the ultimate Soviet status symbol, because only three people in the entire country would get one: the first secretary of the Central Committee, the chairman of the Supreme Soviet Presidium, and the chairman of the Council of Ministers. Like Khrushchev's new house on Lenin Hills, with its gardens and fountains, cherry trees, and panoramic views of the Moskva River, the ZIL was a coveted perk that would help the CIA divine the shifting pecking order of communism's quarrelsome high priests. In the opaque Soviet system, one glance at the Kremlin motor pool could yield more intelligence than a year's subscription to the newspaper *Pravda*.

Pravda never made mention of the work in Kaliningrad, for if the name was even spoken "it was always in a whisper," Sergei Khrushchev recalled. Such was the secrecy surrounding the missile complex that the name of the man who awaited the Presidium delegation this February morning had also been erased from all records. He had been given a pseudonym and was obliquely referred to in official communiqués simply as the Chief Designer. The Chief Designer was a prisoner of his own success. He was deemed so important to national security that a KGB detail watched his every step. For as long as he lived, he would not be permitted to travel abroad. He could not openly wear the numerous medals and citations he received, nor could he have his photograph publicly taken. All this was for his own protection, because the Soviets were convinced that the CIA would try to kidnap or assassinate him. The veil of anonymity would be lifted only after his death, when his name would replace Kaliningrad on post-Communist maps. But back in 1956, Sergei Korolev's true identity was a closely guarded state secret.

Korolev greeted his guests formally, exchanging rigid handshakes with the Kremlin cardinals. Molotov, the aging diplomat, whose signature had adorned the nonaggression pact with the Nazis, was the senior member of the group. By dint of his decades as foreign minister, he was also the most worldly and urbane. Kaganovich, Khrushchev's mustachioed former mentor, was another story. Once rakishly handsome, he had grown fat, old, and ugly, and was resentful at heaving been leapfrogged by his protégé. Kirichenko was the new boy, tall and big-boned. He'd been brought in by Khrushchev, who was trying to pack the Presidium with his own acolytes. The last to extend a hand was Bulganin, Khrushchev's bitter rival, the sly Soviet premier with a sinister silver goatee.

Bulganin and Khrushchev shared power under an uneasy arrangement that satisfied neither man and sowed confusion both at home and abroad. Bulganin, as chairman of the Council of Ministers, headed the government. Khrushchev headed the party. But who was the head of state? According to communist dogma, the government was supposed to answer to the party. But in practice, the subordination was not always so clear-cut. When Eisenhower met both men in Geneva for talks the year before, the president's advisers had spent much of that summit trying to figure out who was really running the show in the Soviet Union. Such an ambiguous arrangement could not last indefinitely. Nor would it.

While the introductions were made, and aides scurried attentively to take their bosses' overcoats and homburg hats, the younger Khrushchev soaked in the surroundings. Buildings of all sizes dotted the missile complex: dark, grime-covered brick structures, huge rusty hangars, water towers, military-style barracks, and corrugated steel sheds that stored, among other things, 1,500 tons of potatoes and 500 tons of cabbage so that NII-88's cantina would not suffer from the food shortages that plagued the rest of the country. There was a decrepit, Dickensian feel to the place, so much so that Sergei Khrushchev mistakenly dated the facility's original construction to the nineteenth century. In fact, it had been built in 1926 by the German firm of Rhein-Metall Borsig to manufacture precision machinery and was later retooled and expanded to produce artillery pieces for the war. Buildings had a tendency to age prematurely under Soviet care.

Inside, the installations were surprisingly clean and modern and gleamed with white paint. The delegation was ushered into one of the largest of these, a brightly lit hangar of imposing dimensions. At the center of the hangar, displayed on large holding rings like precious museum exhibits, lay three rockets. "This is our past," said Korolev, pointing to the smallest of the reclined missiles. Korolev was a short, powerfully built man, with a muscular neck and the compact frame of a middleweight wrestler. He had thick black hair, slightly graying on the sides, which he slicked straight back over his large forehead with the aid of pomade. He spoke slowly, in a tone that was neither obsequious nor insecure. Korolev was accustomed to dealing with Presidium members; he had even reported to Stalin on several occasions after returning from Germany in 1946 and being named head of the newly created OKB Special Design Bureau at NII-88.

The rocket he pointed to was the fruit of OKB-1's German labors, an identical replica of the V-2 called the R-1. Everything about it was German: the parts that had gone into it; the engineers and technicians who had been forcibly relocated to Russia to assemble it (it was from them that the CIA eventually gleaned most of its information on the missile complex); even the camouflage scheme, which mimicked that of the V-2. The R-1, Korolev explained, had taken three painstaking years to master. Not until 1948 had Korolev felt confident enough to try to launch it. It flew a few hundred miles—in the wrong direction.

Khrushchev listened attentively, nodding politely. Much of this he already knew. "Father was no longer a novice when it came to missiles," Sergei Khrushchev recalled. That certainly had not been the case when Beria was alive. Beria, much like Hitler's secret police chief, Heinrich Himmler (whom he even resembled in an effete, murderous way), had tried to dominate his country's missile programs. Though Beria had not exercised remotely the same degree of control over NII-88 as the SS chief had exerted over V-2 production, virtually all top-level decisions involving Soviet missile development had been made by him and Stalin alone, without the participation of other Presidium members. "We were technological ignoramuses," the elder Khrushchev recounted in his memoirs, describing the first time he and his fellow Presidium members saw a missile after Stalin's death in 1953. "We gawked," he wrote, "as if we were a bunch of sheep. We were like peasants in the marketplace. We walked around and around the rocket, touching it, tapping it to see if it was sturdy enough—we did everything but lick it to see how it tasted."

The missile they had seen back then was the R-2, and this was the next exhibit in Korolev's tour of OKB-1. The delegation—Kremlin dignitaries in their somber, medal-bedecked suits, engineers in white smocks, security men from the KGB's Ninth Directorate in black knee-length leather jackets—obediently walked over to the full-scale R-2 model. It closely resembled the R-1, except that it was nine feet longer and of a slightly wider girth, which allowed it to carry extra fuel, doubling its range to nearly 400 miles. The R-2, Korolev explained, was a hybrid: a half-Russian, half-German elongation of the original V-2. It could go farther and faster, climb higher, and carry a heavier payload than its predecessor. Alas, it could not land much more accurately, despite all the 15,000-ruble bonuses offered to captive German engineers to improve the gyroscopic and radio beacon guidance systems that had been developed in Germany during the last year of the war. Still, the R-2 represented a major technological leap in the field of structural design, where the Soviets had learned how to build a much stronger rocket, able to withstand far greater stress loads, without significantly increasing the width and weight of the materials used to build it.

At this, the Presidium members also nodded knowingly. But Molotov and Kaganovich seemed distracted, while Bulganin appeared lost in his own private universe. Perhaps they had also heard the lecture before, or perhaps they were still mulling over the consequences of Khrushchev's secret speech at the Twentieth Party Congress. Barely two days had passed since the dramatic address had so shocked and staggered its audience that, in the words of one participant, "you could hear a fly buzzing" in the stunned silence that permeated the Great Kremlin Palace. Khrushchev had spoken for nearly four uninterrupted hours, and his listeners had turned either deathly pale or beet-red with anger. When he was done, there was not the usual standing ovation, which was typically thunderous and prolonged, because in Stalin's day no one dared to be the first to stop clapping and sit down. The cheering could last five palm-aching minutes until the "boss" signaled that he was satisfied. But Khrushchev, at the Twentieth Party Congress, had done the unimaginable: he had publicly attacked Joseph Stalin. Stalin, he said, had abused his authority and had ruined millions of innocent lives. Stalin was a murderer, a liar, and a thief, who had stolen the communist ideal and perverted it with his paranoid quest for power.

An audible gasp had echoed throughout the palace hall, as if there suddenly was not enough air to breathe. Even to think what Khrushchev had uttered aloud was considered heretical in the Soviet Union, punishable by excommunication to Siberia or outright execution. What Khrushchev had done was tantamount to the pope assembling the College of Cardinals to denounce Christ. Stalin had been a demigod, as much revered as feared. At his funeral, such was the outpouring of grief that one hundred people had been trampled to death by the mobs of mourners that had descended on Red Square to view his body lying in state next to Lenin. His victims sent tearful tributes from prisons. Grown men sobbed inconsolably on the subway. Only the Presidium members shed crocodile tears. As the Great Leader slipped in and out of consciousness, they had denied him medical attention to hasten his demise. They had hovered over his deathbed like ghouls, falling to their knees when his eyes flickered open, retreating to scheme in dark corners when they closed. When at last he drew his final strained breath, the relief on their faces was clear. They had survived the Terror.

But now Khrushchev's speech had put them all in danger again. They had warned him not to do it; it would open up a can of worms that might consume them all, they said. "If now, at the fountain of communist wisdom, a new course is set which appears to deviate considerably from that of the Stalin era," the CIA noted in an April 1956 analysis of the speech, "repercussions are likely to occur which may be of great moment. . . . It may set in motion forces extending far beyond the contemplation of the collective leaders of the CPSU."

Khrushchev told his fellow Presidium members that 7 million Soviet citizens would soon be returning from Stalin's Siberian gulags. From the prisons they would bring back horror stories of torture and mass murder, of starvation and sham trials. There would be no way, Khrushchev reasoned, to keep the full and terrible extent of Stalin's purges quiet much longer.

"Don't you see what will happen?" Kaganovich had protested. "They'll hold us accountable. . . . We were in the leadership, and if we didn't know, that's our problem, but we're still responsible for everything."

Kaganovich's feigned ignorance must have provoked a cynical snicker from the other members of the Presidium. How any Soviet citizen, least of all a fawning Stalin henchman like Lazar Kaganovich, could profess no knowledge of the fratricidal inferno that raged between 1929 and 1953 and engulfed 18 million lives was baffling. What had begun as a settling of accounts in the Central Committee had degenerated into a national feeding frenzy that left no corner of the Soviet empire unscarred. People turned in their neighbors because they wanted their apartments. Subordinates ratted out superiors because they wanted their jobs. Economic failures were blamed on "wreckers" and foreign saboteurs. Ethnic minorities were arrested en masse. Balts, Jews, Chechens, Volga Germans, Crimean Tatars, and Western Ukrainians were shipped by the trainload to Arctic jails. The Siberian death trains ran from as far west as Warsaw, where more than one million Poles were sent to the gulags, half never to return. The gulag archipelago that became their final resting place stretched over thirteen time zones, encompassed three thousand prison and labor camps, and employed hundreds of thousands of guards, administrators, and factory technicians. Vast railway networks had been carved over the permafrost to supply it with inmates and to haul the gold, coal, diamonds, and timber that their slave labor produced. Entire swaths of the Soviet economy depended on the blood-soaked revenue generated by the gulags, which did double duty as an engine of mass murder no less efficient than the Nazi concentration camps. The weeks-long cattle-car journeys could kill half the arrivals. Cold, disease, malnutrition, and overwork did the rest.

It was impossible for Kaganovich not to know any of this, just as it was impossible for the delegates of the Twentieth Party Congress not to have noticed that more than 1,000 of their 1,500 predecessors from the Seventeenth Congress in 1949 had been shot. Of course Kaganovich had known. Everyone knew. Each Presidium member, Khrushchev included, had personally signed thousands of arrest and death warrants. In their speeches, they had all exhorted the security men to exceed their brutal interrogation quotas, to root out shirkers, reactionary right-leaners, and foreign spies. Molotov, in his memoirs, even justified arresting the wives and children of enemies of the people: "They had to be isolated. . . . Otherwise, of course, they would have spread all sorts of complaints and demoralization." Molotov's own wife fell victim to the purges, but he did not try to save her from a Siberian labor camp. Nor did Kaganovich rise to the defense of his brother, who committed suicide rather than face trial. They had had to remain silent, to be more Stalinist than Stalin himself, because betraying even the slightest hint of hesitation would have cost them their lives. "All it took was an instant," Khrushchev recalled. "All you did was blink and the door would open and you'd find yourself in Lubyanka," the KGB headquarters.

For the men of Stalin's inner circle, participation in the purges had been a matter of personal survival. But now that Khrushchev had unmasked Stalin's reign of terror, would they survive an accounting for their actions? That was the real danger of Khrushchev's speech—not the professed knowledge of the purges but the official acknowledgment of them, as the CIA report on the secret speech underscored. "A change from violence to diplomacy and from tension to relaxation cannot but have a deep psychological impact on the people inside the communist orbit" was the analysis in Washington. "The question arises whether the leaders of the CPSU can dispense with permanent tension without at the same time undermining their monolithic dictatorship."

For the first time, a Soviet leader had admitted that horrible crimes had been committed in the name of communism, that terror as a tool of the state was wrong. It was still too early to tell what that meant. But as the CIA predicted, there were almost certainly going to be dire consequences.

These, and other, thoughts troubled Korolev's distracted guests. The Chief Designer would not have known what weighty matters preoccupied his visitors because he had not been privy to the secret speech. But he was intimately acquainted with its contents. He had been among the 1,548,366 people arrested in 1938, an ordeal that began, as in the case of countless others, with a knock at the door in the middle of the night and nearly always ended in death, either by outright execution or later in the camps. "He was taken to Lefortovo prison, interrogated, beaten," a colleague of Korolev's would later inform the biographer James Harford. "He remembered asking for a glass of water from one [guard] who handed him the glass and then hit him in the head with the water jug. He was called an enemy of the people."

Korolev didn't need Khrushchev to tell him how cold it could get in Kolyma, or what it was like to sleep barefoot in the snow with a broken jaw. He had not needed to be told how it felt to lose all your teeth from malnutrition. And he didn't need to listen to a speech to know that the prosecutions had been shams. In fact, the scientist at his side during most of the Presidium visit, the propulsion expert Valentin Glushko, had provided testimony against him at his own trial, which had ended with a ten-year sentence "for crimes in the field of a new technology" and a perfunctory "Next."

What Korolev did need from Khrushchev that day was the green light to embark on an ambitious new project. So the Chief Designer smiled his pearly denture smile at the men who had robbed years from his life and continued his lecture. The first full-scale R-2, he explained, had been field-tested in October 1950. The launch was not a success. After tinkering throughout the harsh winter, the designers attempted another series of test firings in the spring of 1951. This time it flew 390 miles—in the right direction. "Which brings us to the present," said Korolev, gesturing toward the largest of the three missiles. "This is the R-5," he said, tapping the rocket with a pointer, "the first Soviet strategic rocket," built entirely by Russians without German help.

The R-5, the Presidium members could immediately see, was radically different from its predecessors. It was longer by half than the original V-2 replica, slimmer, more fragile in appearance, and tubular in shape. The four big, graceful stabilizer fins that had made the tapered V-2 instantly recognizable had been lopped back to tiny triangle wedges on the R-5, and the nose cone had been blunted into an ungainly snout.

"The construction looked utterly incapable of flight," Sergei Khrushchev recalled, a sixty-six-foot-long pencil in need of sharpening, with an engine for a rubber eraser. "The [R-1 and R-2] at least had streamlined shapes and a certain refinement in the form of stabilizers," he would later write. "Apparently I wasn't the only one to have this reaction, since Father looked surprised."

Korolev, at last, had his guests' undivided attention. Not only did the R-5 fly, he explained, but it spent most of its flight above the atmosphere. Large stabilizer fins were not necessary because of servomotors for the small aerodynamic rudders. The rocket's tubular shell doubled as the propellant tank wall, further reducing design mass by a full ton while increasing fuel capacity by 60 percent. The engine, an RD-103 designed by Glushko, also produced 60 percent more thrust than its predecessors, and thanks to the introduction of coolant flowing through integrated solder-welded ribs around the combustion chamber, it could operate longer and more efficiently, without overheating or cracking under the intense temperatures and pressure generated by forty-four tons of thrust. Glushko's engine propelled the R-5 to a top speed of 10,000 feet per second, or twice as fast as the V-2. A new thermal shield protected the nose cone from the excess heat generated by the increased velocity and atmospheric reentry, while targeting accuracy had been greatly improved with the addition of longitudinal acceleration integrators, which could control engine cutoff with greater precision.

Fully fueled and armed, the R-5 weighed twenty-nine tons. Its range was 1,200 kilometers, roughly 800 miles. Its payload was an eighty-kiloton nuclear warhead, the equivalent force of six Hiroshima bombs. And the R-5 was not a mock-up or test vehicle. It was operational. Three weeks earlier, on February 2, the missile had carried its lethal cargo 800 miles, setting off a mushroom cloud over a target area near the Aral Sea in Soviet central Asia. The test had marked the world's first nuclear detonation delivered by a ballistic missile, the dawn of a new age in warfare. The Soviet Union had fired the first salvo of an arms race that would consume trillions of dollars and hold the planet hostage for the next forty years.

The Presidium members stared intently at their revolutionary new weapon. It seemed incomprehensible that such a strange, fragile object could wield such power; that with one push of a button it could vaporize an entire city in an instant. Khrushchev and a few of the other Presidium members had seen war, and they knew that it was incremental, a process of attrition. In the sieges of Stalingrad and Kursk, Leningrad and Kiev, the devastation had been progressive. A little bit of each city had died each day, and the process had lasted for agonizing months. With the R-5, everything would be over within seven minutes of launch. You didn't need planes, tanks, or troops, or an invasion fleet. You didn't need to worry about logistics or supply trains. You didn't need to put your soldiers in harm's way. It made war seem almost effortless.

For nearly a minute no one spoke. Khrushchev finally broke the silence. Which countries were in its range? he wanted to know. His son recorded the scene: "Korolev walked over to a map of Europe, which was hanging on a special stand. It looked just like the ones we had had in school, except that this one had arcs of intersecting circles against the blue background of the Atlantic Ocean. Thin radii drawn with India ink stretched to [the Soviet bloc's] western borders, to the frontier of East Germany. In the upper right-hand corner of the map there was a calligraphic inscription: 'Highly classified. Of special importance.' Slightly below that was Copy Number—I don't remember the number, but it was no higher than three."

The map showed that the R-5 could strike every nation in Europe, except Spain and Portugal, which were still out of range. A murmur of satisfaction rose from the Presidium members. "Excellent," said Khrushchev. "Until recently we couldn't even dream of such a thing. But the appetite grows by what it feeds on. Comrade Korolev, isn't it possible to extend the rocket's range?"

No, the Chief Designer replied flatly, a new missile would be required. Khrushchev and the others seemed disappointed. But Korolev appeared unperturbed; his authority was already well established within Kremlin circles. The delegation remained rooted in front of the map for some time, contemplating Armageddon. "Father stared piercingly at it," Sergei Khrushchev recalled.

"How many warheads would be needed to destroy England?" the party leader finally asked. "Have you calculated that?"

Dmitri Ustinov, the armaments minister, fielded the question. "Five. A few more for France—seven or nine, depending on the choice of targets."

Only five? Khrushchev seemed skeptical. The British had withstood a daily barrage of V-2s. They had shrugged off the Junker bombers during the Blitz, displayed a tenacity that even Stalin had praised. But now Great Britain was America's closest ally; it would have to be taken out first.

"Five would be enough to crush defenses and disrupt communications and transportation, not to mention the destruction of major cities," Ustinov explained. His tone, Sergei Khrushchev remembered, "did not allow for even a shadow of a doubt."

"Terrible," said Khrushchev, trailing off in thought. Five flicks of a switch to break the will of a nation like Britain. Astonishing. Cost-effective, too.

• • •

It was not only the search for a new weapon to counter American air superiority that had brought Khrushchev to Korolev's design bureau. He was also looking for ways to save money because, unlike the booming American economy, Soviet central planning was in trouble. The years of rapid growth after the war, when entire cities and industries had been rebuilt from rubble, were over by the mid-1950s. Most Western European nations had made remarkable recoveries by then, due in part to the Marshall Plan. But the command economies of the Eastern bloc had spurned American financial aid and were now beginning to suffer from a crisis of inefficiency. Put simply, Soviet central planning was suffering from the law of diminishing returns. No matter how many rubles the Russians sank into their wobbly industries, they were getting a smaller and smaller return on their investment. And the problem was only going to get worse over time.

For Khrushchev, the economic slowdown came just as his treasury faced the twin challenge of meeting the raised American threat and rising consumer demands at home. After years of sacrifice, first during the war and then during reconstruction, Soviet citizens were beginning to yearn for a higher standard of living. Encouraged in part by a slew of lighthearted films and musicals after Stalin's death that featured fashionable clothes, fast cars, and the joys of shopping, Soviet citizens had embraced this new material slant on life. Khrushchev had approved the cinematic thaw and the departure from the austere militarist offerings of Stalinist directors because he understood that he could not rule by fear alone. Stalin had shrewdly substituted a diet of national pride and terror for material well-being that had given his subjects comfort in the notion that while they were frightened and poor, they were building a great empire and a better tomorrow. Khrushchev preferred the carrot to Stalin's stick. Parades were not enough. People had to have tangible rewards from the socialist miracle touted by the new propaganda films.

To bring his movies more in line with reality, Khrushchev needed to free up hundreds of billions of rubles to pay for housing projects, to re-capitalize decrepit automobile factories, and to finance agricultural reforms that would put food on people's tables. As things stood, the Soviet Union could barely feed itself. Yields were so low that in 1953 per capita grain production had fallen to 1913 levels. All told, the USSR's muddy collective farms were producing less food in the early 1950s than they had in 1940. If the trend continued, the Soviet Union would starve.

To reverse the decline, Khrushchev had embarked on a hugely ambitious and controversial program to develop 80 million acres of virgin steppe in central Asia that was intended to increase agricultural production by 50 percent. Molotov, in particular, had vehemently opposed the plan, which required relocating three hundred thousand farmworkers and fifty thousand tractors to cultivate the raw fields in Kazakhstan and southern Siberia. It was far too expensive, Molotov argued, and the climate was too harsh. But Khrushchev had overruled him; the money, he said, would be found somewhere, trimmed from the fat of various other budget allocations.

Military expenditures ate up between 14 and 20 percent of the Soviet economy, compared to 9 percent for the United States, whose economy was much larger. And yet, despite the large outlays, the Soviet armed forces could not adequately address the new American jet-bomber threat. The Red Army had more than 3.5 million men under arms, but they were useless against B-47s carrying nuclear bombs. What's more, they were expensive; soldiers had to be clothed, fed, housed, and provided with costly trucks, tanks, and artillery whose role in a nuclear standoff was of limited or no value. Khrushchev could try to compete with the Americans by bulking up his aging fleet of World War II–era bombers, but it was becoming clear that he could not keep pace with the U.S. Air Force. Even if Khrushchev ramped up spending, he could not hope to match the new B-52 superbomber that Boeing was beginning to mass-produce. In heavy aviation, the Soviet Union had simply fallen too far behind to make the investment worthwhile.

In the high-stakes arms race, Khrushchev was a pauper playing at a rich's man table. Given the vast financial gulf between America and the Soviet Union, he had to marshal his resources more carefully than Eisenhower, and he saw that missiles were the cheapest, most cost-effective way to stay in the game. Their costs were mostly up-front, in research and development. And Khrushchev felt that he would not need to produce them in significant numbers. That hadn't been the case in World War II. Hitler, after all, had bet the Reich on the V-2, built thousands of them, and lost. But that was before the atomic age. Nuclear warheads dramatically changed the equation. Now you could defeat England with only five rockets and keep America at bay with a few dozen more. From a financial point of view, focusing on missiles made more sense than maintaining a huge standing army or sinking money into giant bomber fleets that rockets would soon render obsolete anyway.

And so Khrushchev was betting the farm on Korolev. He had slashed spending on conventional forces to free up funds for foodstuffs and housing, while he hoped that the Chief Designer's missiles would provide the USSR with a shield against American bombers. Just how big a wager he was placing would stun the CIA, which in a 1958 report underscored "the striking re-allocation of expenditures within the [Soviet military] mission structure. The most dramatic examples are the 34% decline in expenditures for the ground mission, and the 127% increase for the strategic attack mission. . . . Increasing expenditures on strategic attack reflect the replacement of the manned bomber by long range missiles."

Khrushchev, in short, was gambling that the Americans had put their money on the wrong weapons system.

· · ·

"Well," asked Khrushchev, "what else do you have to show us?" This was the moment Sergei Korolev had been waiting for; the moment he had prepared for ever since the day he was released from prison and sent to Germany to join Chertok and Glushko to uncover the secrets of Nazi missiles. (Chertok, alas, could not share in the triumph. He had been demoted from his post as NII-88 deputy director during one of Stalin's anti-Semitic roundups. After Stalin's death, Korolev eased Chertok back into a senior position at OKB-1, but his rank was not high enough to meet Presidium members.) Now NII-88's most senior scientists stirred anxiously as Korolev prepared to unveil their collective masterpiece, the product of over a decade of research and millions of man-hours of dedicated work.

"We have seen the past and the present of Soviet rocketry," said Korolev, leading his guests to a large double door, where a pair of crisply uniformed guards stood rigidly at attention. "This is the future."

The doors swung dramatically open, and Sergei Khrushchev gasped. "I was amazed. I had never seen anything remotely like it, no one had," he said, the excitement and wonder of that first glimpse still evident even after fifty years. Sergei and his father stepped into the top-secret inner room. The structure was brand-new, so spotless that everything shone, and its walls soared upward instead of lengthwise like the hangar they had just come from. These walls were constructed entirely of glass, and they had been painted white to allow light in but to keep prying eyes out. At the center of the gigantic atrium stood a rocket larger than Sergei Khrushchev had ever imagined. "It looked like the Kremlin tower," he said.

Korolev stood aside and savored the moment. The rulers of the Soviet empire, among the most ruthless and powerful men on the planet, were frozen with awe. They had stopped dead in their tracks, reduced momentarily to the status of mere mortals. "Father later told me that he was simply numb, intimidated by the grandeur of such an object created by human hands," Sergei recalled.

"The R-7," announced Korolev with theatrical flair. The Presidium members recovered some of their composure and slowly circled the mammoth missile. In weight and mass it was ten times the size of the R-5, and almost twice as heavy as the Hindenburg Zeppelin, the largest aircraft ever built. Whereas the R-5 had one engine, the R-7 boasted five giant boosters that would consume 247 tons of fuel in four minutes. The nearly one million pounds of lift they generated could hurl the missile more than 5,000 miles at a speed of over 24,000 feet per second, four times faster and forty times farther than the original V-2.

The Presidium members drank this all in like a golden elixir. They poked their bald pates into the missile's twenty burnished copper exhaust nozzles and craned their creaky old necks up at the inky black nose cone where the thermonuclear warhead would sit, a five-ton device that would have an explosive yield nearly one hundred times that of the atomic bomb the Americans had dropped on Hiroshima. The men of the Presidium shook their heads in wonder, whistled softly, and shot one another glances that brimmed with satisfaction. At last the Soviet Union would have its ultimate weapon, a rocket that could reach New York and Washington with the deadly force of all the combined ordnance dropped in the Second World War.

But try as they did, the Presidium members could not entirely fathom the full lethal potential of the R-7. Just how fast is 24,000 feet per second? Khrushchev tried to imagine.

"How long would it take to fly to Kiev?" he asked Korolev. When he was Stalin's Ukrainian viceroy, Khrushchev had made the arduous three-hour air commute between Moscow and the Ukrainian capital several times a week in his old Douglass.

"Maybe a minute," Korolev replied.

Khrushchev whistled appreciatively. Bulganin stroked his goatee. That meant the R-7 could reach the United States in less than half an hour. Even if the radar stations the Americans had built in Norway, Turkey, and Iran picked up the launch, they would not have enough time to react, to scramble the Strategic Air Command or evacuate their cities. The R-7 really could make all the difference, change the entire security dynamic—if it worked. And that would not be known until later that year, when testing could begin.

"Why is the rocket tapered in the middle?" Molotov asked, interrupting Khrushchev's train of thought. The R-7 had a slightly hourglass-like shape because the four big boosters strapped around the central core flared out like a skirt. Korolev began to explain that the narrow midriff was necessary for the strapped-on engines to attach and jettison properly, but Khrushchev snapped impatiently: "Why do you ask such stupid questions? You don't understand anything about these technical matters. Leave them to the experts."

Molotov shrank back at the public rebuke but said nothing. There was a time when Khrushchev would never have dared to humiliate him so openly, when Khrushchev had fawned over him and genuinely admired him. But now the upstart was putting on airs, interfering in foreign affairs—even though he had never traveled abroad until after Stalin's death. Once, when they had been coconspirators, Khrushchev had known his place. But now he was becoming impossible, an expert on everything. Behind his back Molotov called Khrushchev a "small-time cattle dealer. Without a doubt a man of little culture. A cattle trader. A man who sells livestock." Khrushchev, on the other hand, derided the old diplomat as a geriatric Stalinist relic. "Their relations had become tense," Sergei Khrushchev recalled, "especially after the secret speech."

Like Kaganovich, Molotov had been vehemently opposed to Khrushchev's de-Stalinization program. For the foreign minister, there were external considerations. Fear of Stalin had been the glue that cemented the Eastern Bloc. After his death, rumblings of discontent had begun to ripple throughout the captive states of central Europe, particularly in Poland and Hungary. In Yugoslavia, Marshal Josip Broz Tito had become downright disrespectful, and Molotov had been furious that Khrushchev had not punished him. It set a bad precedent to be seen as so soft. And now that Khrushchev had denounced Stalin's reign of terror, the Poles and Hungarians might also grow bolder. Khrushchev's speech could be taken as a sign of weakness throughout the Soviet dominions, a reflection of waning resolve, a cue to rise up. Khrushchev, the novice, didn't understand any of these things; he had no comprehension of the forces he might have unleashed. His pigheaded ignorance could bring down the whole empire.

Molotov, of course, said none of this publicly. He was too seasoned a Kremlin intriguer to make that mistake. Publicly he recanted his opposition to appeasing Tito, saying, "I consider the Presidium has correctly pointed out the error of my position," and joined Kaganovich in praising Khrushchev on his insightful initiatives. "Comrade Khrushchev carries out his work . . . intensively, steadfastly, actively and enterprisingly, as befits a Leninist Bolshevik," he had said only a few months earlier. But privately, Kaganovich and Molotov were already whispering in Bulganin's malevolent ear. Now was not yet the time. Like the butcher Beria before him, the cattle salesman would get his comeuppance soon enough. With luck he might not even live to see the R-7 fly.

• • •

"I would like you to know about still another project," Korolev said quickly, as the Presidium delegation was about to leave. For the first time a hint of hesitation had crept into the Chief Designer's normally self-confident tone, and his words had gushed out in a torrent of pent-up anxiety.

Korolev "led us to a stand occupying a modest place in the corner," Sergei Khrushchev recalled. "A model of some kind of apparatus lay on the stand. It looked unusual, to put it mildly. A flying machine should have a smooth surface, flowing shapes and clean-cut angles. But this one had some type of rods protruding on all sides and paneling swollen by projections."

What is it? the Presidium members asked. A satellite, said Korolev. He paused for effect, gauging his guests' reaction. There was none. Instead, they stared blankly at the meaningless object. Korolev must have sensed the disinterest, for he launched into an impassioned speech. From time immemorial, he said, growing animated and uncharacteristically emotional, man has dreamed of escaping the bonds of gravity, of breaking free of the earth's atmosphere and exploring the cosmos. Until now the dream of the space pioneers—and here Korolev spoke glowingly of the nineteenth-century Russian rocket visionary Konstantin Tsiolkowsky—had belonged to the realm of theory or science fiction because no man-made object could generate sufficient velocity to break the gravity barrier. The R-7, though, was almost fast enough. With a little tinkering and a few minor adjustments, it could make that age-old dream possible.

Once again, Korolev paused and looked at his guests. The Presidium members seemed unmoved. So what? their expressionless faces seemed to say. What did any of this have to do with the development of an intercontinental ballistic missile that could keep the Soviet Union safe from American attack? How could the two even compare in national importance? Korolev was wasting their time with this romantic nonsense. The Chief Designer had been getting this sort of blasé reaction for two years now, ever since his proposal for a satellite project had begun wending its way slowly up the Soviet bureaucracy from one skeptical committee to the next. Decrees had been signed advocating the "artificial moon" as far back as May 1954, but without a champion on the Presidium to lend weight to the resolutions they were just pieces of paper. "You needed the constant support of power," Sergei Khrushchev explained of the way things worked in the dictatorship of the proletariat. "Everyone needed to know that you could pick up the phone and dial the First Secretary's four-digit extension number directly if there was an obstacle. Otherwise you would fail."

Korolev must have sensed that he was losing his audience, and his one chance to get Khrushchev or one of the others to personally sign off on his pet project. He quickly changed tack. The Americans, he said casually, were in the advanced stages of developing a similar satellite. This was a slight exaggeration, but the Presidium didn't need to know that. The United States, Korolev continued his pitch, had been working on a satellite for some years. He was certain, though, that he could beat them to space. It would be a significant scientific victory, he added, not to mention a serious defeat for the capitalists. Bulganin and Molotov looked at the model satellite with renewed interest. The shadow of a smile formed under Kaganovich's dark mustache.

Korolev's ploy had not been subtle, but it had its desired effect. He decided to press his advantage: "The Americans have taken a wrong turn. They are developing a special rocket and spending millions. We only have to remove the thermonuclear warhead and put a satellite in its place. And that's all."

Sergei Khrushchev recalled his father staring long and hard at the satellite model, mulling over Korolev's request. "It seemed as if he was still debating the matter," he observed. Part of the problem was that other than the prestige of being first in orbit, the satellite didn't appear to have much of a purpose. Korolev had spoken of scientific readings and radio signals, but the men of the Presidium failed to see the point. They were not alone. Only a few hundred people on the entire planet in 1956 grasped the true potential and significance of a satellite, and several of them happened to work in surveillance at the CIA. For the leaders of the Soviet Union, dreams of distant space conquests risked becoming costly distractions from the immediate and earthly concerns of the cold war.

Minutes passed, and Korolev once again assured his masters that launching a satellite would in no way interfere with the development of the ICBM since it could only occur once the R-7 was fully operational anyway. Khrushchev seemed to weigh this. The ICBM was unquestionably the Soviet Union's overriding priority. But the prospect of thumbing his nose at the arrogant Americans also had a certain undeniable appeal.

Okay, he finally relented. "If the main task doesn't suffer, do it."

2

JET POWER

Five thousand miles from Moscow—beyond bomber range—the Alabama sky brimmed a slate blue. To the west rose the Appalachian foothills, rolling like a brown carpet of dead leaves. The Tennessee River ran south, also brown and undulating, a giant corn snake weaving through the fallow cotton fields. To the east stood Huntsville, ancient Confederate battleground and the new home of the U.S. Army Ballistic Missile Agency.

The plane was coming in from the north, from Washington, whence the money and decisions flowed. Major General John Bruce Medaris stood on the tarmac, scanning the wintry horizon, searching for the contrail that would announce the impending arrival of the secretary of defense. Pacing at the edge of the runway, Medaris flicked his riding crop impatiently. A memento from his days serving with General George S. Patton, the swagger stick, along with his slender mustache, hazel eyes, and vaguely piratical air, lent Medaris a remarkable resemblance to Errol Flynn. This he knew, for he was vain, and his vanity, with its attending indiscretions, had already cost him one marriage— and very nearly another.

Dashing was the term the newspapers used to describe him. *Belligerent*, *abrasive*, and "a troublemaker who was hard to handle" were a few of the other, less flattering descriptions of Medaris, who would bluntly reply that "politeness is nice but takes too damn much time." There was no middle ground with the fifty-three-year-old general. "You either loved him or hated his guts," in the words of one subordinate. Those who served under Medaris tended to fall into the former category. Those he answered to in the Pentagon were usually in the latter.

And now they were coming on an inspection tour, these politicians and pencil pushers who were the perennial scourge of field officers. As a career Ordnance man, Medaris did not storm beaches, but he supplied the munitions for those who did. During World War II, he had moved thirty thousand vehicles for General Patton in the North African and Sicilian campaigns, and he had equipped General Omar Bradley's entire First Army Group for the D-Day invasion. These were impressive logistical feats, but not the stuff of glory that made the newsreels. Still, Medaris managed to win a medal for bravery at Omaha Beach, in addition to numerous other combat citations and awards, and in spirit he identified more with the hard-charging fighting men of old than with the cautious new breed of technocrats taking over the military. Unfortunately for his career, this shared affinity included a disdain for authority and an enduring allergy to regulations.

It wasn't that Medaris didn't respect rules. But like Douglas MacArthur and Patton, he simply didn't think they applied to him, as the Huntsville military police had discovered a few weeks earlier when he roared into his new command in a Jaguar and dressed in golf attire. "Didn't you see the speed limit sign back there?" the startled MPs demanded.

"What did it say?"

"Forty-five miles an hour and you were going sixty."

"Son, I'm General Medaris and the speed limit is now sixty."

It was in this characteristically rebellious manner that Medaris assumed command of the newly created Army Ballistic Missile Agency on February 1, 1956. ABMA had just been founded by administrative fiat in what was essentially a bureaucratic counteroffensive by the army to keep the burgeoning air force in check. The old Army Air Corps, once an insignificant asterisk in the army's accounting ledgers, had become an unstoppable juggernaut since gaining its independence as a separate service in 1947. In the nuclear age, bombers, not tanks, kept America safe, and it was pilots, not grunts, who were the darlings of politicians and policy makers. John Foster Dulles's strategic deterrent so strongly favored the young air force that it now swallowed forty-six cents of every military dollar. Its manpower now nearly equaled that of the army, whose budget and personnel had shrunk by half, and air force assets in 1956 exceeded those of the fifty-five largest U.S. civilian corporations combined.

The army was thus fighting a rearguard action to stay relevant in the rapidly shifting military pecking order. The humiliating infantry debacles of the Korean War had not helped its cause, and missiles offered the West Pointers one of the few new areas with potential for expanding their role. Rockets, after all, were natural extensions of artillery. The air force, however, had different ideas, making the case that missiles were nothing more than delivery systems, effectively unmanned bombers, and thus ought to be assigned to the Strategic Air Command. Not to be left out of the squabbling, the navy quickly developed its own distinct missile doctrine and leaped into the fray as well.

An all-out rocket war had erupted among the three services, and it was into this internecine conflict that Medaris was thrust as the army's point man. Ironically, the very same character traits that rubbed his superiors the wrong way had recommended him for the post. "You are aggressive. Some would say to a fault," he had been told on winning the ABMA job, hardly a customary endorsement. But right now the army needed someone with his particular talents for this difficult mission.

Medaris reflected on his new assignment as he waited for the big military transport bearing the secretary of defense to arrive from Washington. His own plane, a four-seat Aero Commander, sat at the other end of the tarmac, and in the spirit of interservice rivalry he preferred to pilot it himself rather than trust his fortunes to the air force.

There had been little time to prepare for this important visit, and in the few weeks afforded to him, Medaris had done what he could to whip the month-old missile agency and its dilapidated buildings into shape. ABMA's headquarters had been hived off the 40,000-acre Redstone Arsenal, a neglected World War II munitions and chemical weapons plant that did not enjoy "a great reputation at that point," in Medaris's own words. Black skull-and-bones contamination warning signs still hung from rusted barbed-wire fences strung around the skeletal remains of abandoned chemical depots. Squat, circular storage bunkers dotted the landscape like concrete igloos. The cavernous old assembly lines and the cracked and grimy windowpanes gave off an air of postindustrial misery.

ABMA's fortunes seemed nearly as grim as the headquarters the army had given it. But it did have one ace in the hole: Wernher von Braun and the brain trust behind the V-2, the greatest team of rocket scientists on earth. Medaris's first order of business had been to cordon off von Braun's research facilities from the rest of the ramshackle base, bypassing procedure with a scribbled note on the back of an envelope that read "You are authorized to procure and install fencing." This was typical of Medaris and decidedly not the way the Army Quartermaster's Office did things, yet another reason he had been denied promotion from colonel to brigadier general on three consecutive occasions during the war, despite endorsements from Bradley, Patton, and Eisenhower himself. Corner-cutting got timely results for the frontline generals, but it left a lot of noses out of joint back in Washington, where they preferred their paperwork duly filled out in triplicate.

With only a few weeks to prepare for the secretary of defense's inspection tour, Medaris hadn't had time for bureaucratic niceties. He'd ordered the buildings scrubbed and the grounds swept. ABMA's 1,700 civilian scientists had been issued strict instructions to tuck in their shirttails and assume a more military posture. Special flags and insignias had been designed and distributed to impress guests and instill esprit de corps, and MPs in parade uniforms, each man at least six feet tall, had been posted outside doors, elevators, and anywhere else a VIP delegation might venture. Medaris had even refurbished an old plantation log house as a hospitality center to make the secretary's stay more pleasant.

Just about the only thing he had not anticipated was how his well-laid plans would backfire.

• • •

Of all the corporate titans who made up President Eisenhower's "cabinet of millionaires," none was bigger than Secretary of Defense Charles E. Wilson. White-haired and blue-eyed, with a bulldog's bulky frame, the Ohio native had run General Motors with an iron fist since 1941, overseeing its huge defense production during the war and its ambitious retooling efforts afterward to put a car in every American driveway. Under his folksy and forceful stewardship, GM had become a symbol of America's awesome industrial might. Of the 7,920,000 automobiles sold by Detroit in 1955, a 2-million-vehicle increase over 1954, more than half had been built by GM. It was Wilson, a workaholic who usually slept just three hours a night, who had given America the V-8 engine and had fueled the country's passion for size and speed. He had pioneered the monthly car payment plans that made financing the preferred method of purchasing automobiles, and he had tamed the unions by negotiating productivity and cost-of-living escalator clauses that ensured labor peace for decades.

Wilson's skillful planning and execution had made General Motors the biggest company in the world, and he himself had come to personify the new class of American executives democratizing boardrooms across the country. They were midwesterners by and large, from small towns and state colleges, who didn't have Brahmin pedigrees or suites at the opera. Plain-spoken and plain-clothed, they vacationed on the Great Lakes rather than in the south of France and collected hunting rifles instead of antique rugs. Wit, and hard work, not family connections, had gotten them to the top, and the elite East Coast establishment had had no choice but to make room for them. The self-made tycoons were hailed as examples of the new American meritocracy.

For all his success in the private sector, Engine Charlie—thus nicknamed to distinguish him from Charles "Electric Charlie" Wilson, the former General Electric chairman who had run President Harry S. Truman's Office of Defense Mobilization—was not a natural fit as a public servant. Congress took an immediate dislike to him when it emerged during his confirmation hearings that he planned on keeping his GM stock while serving as defense secretary. The government post, he felt, already entailed a significant financial sacrifice in that his salary was diminishing from $566,200 to $22,500. There was no reason to surrender his shareholdings as well. Asked if this might pose a conflict of interest, since GM was one of the Pentagon's largest contractors, Wilson had haughtily replied that he could not conceive of a situation where he would rule against the company anyway because for years he thought "what was good for the country was good for General Motors and vice-versa."

This famous quote would set the tone for Wilson's contentious relationship with the press, which delighted in reporting that Engine Charlie suffered from terminal "foot-in-mouth disease," an affliction stemming from an excess of confidence and wealth. Blunt to the point of profanity, Wilson had no reservations about expressing his views. Congress was a "dung hill"; Democrats were "kennel dogs" that yelped helplessly while waiting to be fed; the National Guard was a "draft-dodging" refuge for cowards that needed to be disbanded; basic research and development was a wasteful scientific boondoggle since it was pointless to "worry about what makes the grass green or why fried potatoes turn brown." His pronouncements on subjects as far ranging as civil rights and agriculture astounded the press and often drove the mild-mannered Eisenhower crazy. "Damn it, how in the hell did a man as shallow as Charlie Wilson ever get to be head of General Motors?" Ike once exploded.

But the president stuck by his outspoken defense chief because military spending was out of control and Engine Charlie, for all his lack of tact, was a ruthless cost cutter. "In his field, he is a competent man," Eisenhower wrote in his diary, shortly after appointing Wilson. Engine Charlie reciprocated by firing forty thousand Pentagon civilian employees in his first few weeks on the job, and by 1956 he had slashed $11 billion from bloated defense budgets. Despite the cuts, military spending still ate up more than half the federal budget, threatening Ike's Great Equation: the delicate balance between a strong economy and a "sufficient" fighting force to best guarantee national security. Engine Charlie's ongoing brief was to locate more fat that could be trimmed, and his rough chopping guidelines were outlined in the New Look Defense Policy—National Security Council document NSC 162/2—that established where America got the biggest bang for its defense buck. Nuclear and bomber programs were inviolate under the New Look doctrine, which rested almost entirely on the buildup of Secretary of State Dulles's "massive retaliatory capabilities." The air force's Strategic Air Command, as the primary instrument of the nuclear deterrent, was off-limits. Everything else in the military kitty was subject to cutbacks, as General Medaris was about to discover.

• • •

Bruce Medaris and Charlie Wilson, by all measures, should have gotten along famously. Both hailed from Ohio, from similarly hardscrabble small industrial towns. They were both practicing Episcopalians, with equally dim views of big government and bureaucracy. Each spoke his mind freely, to his own self-detriment, and Wilson was an abrasive match for Medaris. The two even shared an automotive background. Medaris had worked for the General Motors Export Corporation in South America and had owned a car dealership in Cincinnati before opting for a military career. (He sold Chryslers, but Engine Charlie didn't need to know about that.)

With so much common ground, and ample stores of the salted nuts the defense secretary was said to adore, Medaris thought ABMA well prepared for the inspection tour. And the visit started out promisingly enough. Arriving in Huntsville, Wilson was accompanied by the secretaries of the army and navy, and by General James Gavin, the chief of Army Research and Development, along with several other two- and three-star generals. No air force officials were present in the delegation, owing to the archrivalry that had sprung up as a result of the New Look doctrine, though both ABMA and the Air Force Ballistic Missile Division in Los Angeles had recently exchanged liaison officers—"ambassadors to unfriendly nations," as Medaris wryly described the envoys.

Wilson asked few questions as he toured ABMA facilities at the Redstone Arsenal, viewing the metallurgical and chemical laboratories in buildings 111 and 112; the new static firing stands, where rocket engines were strapped to twin towers and fired at full blast; and the supersonic wind tunnels. Everything had been built for less than $40 million, and Medaris was proud of what von Braun had achieved on a shoestring budget. Inside the main production plant, a hangar where rockets were assembled, von Braun himself showed off the latest batch of his Redstones, America's biggest operational missiles.

Work on the Redstones had begun under the Truman administration, when another automotive industry giant, K. T. Keller, simultaneously served as chairman of the board of the Chrysler Corporation and the Defense Department's director of the Office for Guided Missiles. The Redstone was a heavy-lift tactical missile capable of flinging a 3,500-pound nuclear warhead 200 miles downrange. Essentially a more powerful and advanced version of the V-2, it incorporated a slew of in-novations such as the expanded use of lightweight aluminum alloys, transistors, signal-activated flight steering, and a Rocketdyne engine with 50 percent more thrust. Thirty-seven of the rockets had been built: sixteen in-house at Huntsville and the rest outsourced to Chrysler to "speed production."

Chrysler was also slated to be the prime assembly contractor on the Redstone's successor, the 1,500-mile-range Jupiter missile. Jupiter was the army's bold gamble, the reason ABMA had been created and Medaris dispatched to Huntsville. The air force was already developing a nearly identical midrange missile, the Thor, and had been given sole jurisdiction over the intercontinental arena, with the proposed Atlas ICBM. The programs, fortunately for Medaris, were still in their in-fancy because America had entered the missile sweepstakes so late in the game. The Pentagon brass had felt no need to rush ahead with ex-pensive big rockets because of America's overwhelming advantage in bombers and the overseas bases that placed the Soviet Union within easy reach. As Eisenhower himself noted, the United States already had the proven capacity "to inflict very great, even decisive, damage" on Russia. "The guided missile is, therefore, merely another, or auxiliary, method of delivering the kind of destruction" that Washington already possessed. Besides, Eisenhower argued, he wanted no part in a costly new arms race that would only raise the destructive stakes and plunge the treasury into further debt. "The world in arms is not spending money alone," he had said in 1953, in what was probably the most elo-quent speech of his entire presidency.

> It is spending the sweat of its laborers, the genius of its scientists, the hopes of its children. The cost of one modern heavy bomber is this: a modern brick school in more than sixty cities. It is two electric power plants, each serving a town of sixty thousand popu-lation. It is two fine, fully equipped hospitals. It is some fifty miles of concrete highway.

With 42,000 miles to pave in the massive Interstate Highway System he was proposing, Ike had no interest in trading missiles for schools or bridges, and with little strategic or political incentive to hurry, the air force had spent a mere $14 million developing its ICBM by 1954. Missiles might have languished on the back burner had it not been for the frightening intelligence reports that began trickling in from Moscow the following year. A panel of leading technology experts at the Office of Defense Mobilization led by James Killian, the president of the Massachusetts Institute of Technology, issued a dire warning that unless the United States made immediate and significant strides in missile development, it risked falling hopelessly behind the Soviet Union. The Killian report raised enough alarm that the Eisenhower administration increased missile spending to $550 million in 1955, which still represented a marginal 1 percent of the military budget. That figure was doubled to $1.2 billion in 1956, but that was still far below the $7.5 billion earmarked for beefing up the bomber fleet that year, and was a source of contention within the administration.

Did the accelerated spending "go far enough"? asked Vice President Richard Nixon at a September 8, 1955, meeting of the National Security Council, during which increased missile development was discussed. There could be political repercussions, the vice president worried, if the administration was not seen as doing everything in its power to close the Soviet lead.

"For my conservative blood, it's enough," said Assistant Secretary of Defense Donald Quarles, who also served as secretary of the air force. Quarles was Wilson's closest confidant and protégé, his principal hatchet man. Diminutive, icy, and unfailingly polite, famous for subsisting on seemingly nothing more than cup after cup of steaming hot water, Quarles could afford to cut off the vice president. Nixon was in no position to argue, since Eisenhower had still not decided whether he would ask him to stay on for the 1956 reelection campaign. For close to a year, the young vice president would suffer in limbo, as Ike pointedly refused to rule on his fate, dodging reporters' questions with a humiliating persistency.

Nixon had largely been brought on board the Republican ticket in 1952 because of his youth; just thirty-nine years old, with jet-black hair and a jutting jaw, he projected a vigorous and combative air that counterbalanced Ike's gentle and grandfatherly pate. He had been the youngest member of the Senate, and his selection as Ike's running mate had inspired another youthful legislator, Representative John F. Kennedy, to write a congratulatory note: "I was always convinced that you would move ahead to the top—but I never thought it would come this quickly." Nixon's rise had indeed been meteoric. But perhaps because of the generation gap, he had never hit it off with Eisenhower, who worried about "his lack of maturity."

"You're my boy," the president had exclaimed cheerily upon their first meeting, thus setting the patronizing tone of their unequal union.

With his future uncertain, Nixon bit his tongue as Quarles explained that "the manned aircraft was superior to the ICBM, both in terms of accuracy and weight of destructive force," and said nothing when the assistant secretary closed the discussion by warning that he was opposed "to providing any stronger basis than already existed for individuals . . . who wanted another billion or so" for missile programs. Quarles understood his brief, even if the impetuous vice president apparently didn't. The administration was supposed to slash spending, not encourage it.

Medaris, of course, had not been privy to any of this but was surprised at how uncharacteristically quiet Wilson had been throughout the inspection tour. It was not until the delegation returned to the renovated guesthouse that Engine Charlie grew genuinely animated. "Mr. Wilson started to ask some odd questions," Medaris recalled. "What did it cost to paint those logs so that they would look so pretty?" the defense secretary wanted to know. "He then began asking all sorts of questions about the cost of the quarters, how much money we had put into them, and so on," Medaris said.

Within days of Wilson's return to Washington, Medaris found himself on the receiving end of a series of increasingly persistent queries from the Comptroller's Office at the Pentagon about the farmhouse expenditure.

And so, while Nikita Khrushchev was throwing all his available resources into building the world's first ICBM, the army's top missile commander was kept busy filling out financial spreadsheets in order to prove that the renovation of ABMA's guesthouse would provide taxpayers with a 9 percent return on investment over the comparative cost of renting hotel rooms for visiting dignitaries. For Medaris, "it was the first of many shocks to come."

• • •

While Medaris fended off Wilson's accountants, more bad news for ABMA brewed in Washington. Alarming stories began appearing in the American press of an allegedly huge ramp-up of Soviet heavy bomber production. The USSR, it seemed, was building a fleet of long-range Bisons and Bears at a dramatic rate and would soon be able to launch a first strike against the United States. It was not clear where the media were getting their anonymously sourced information, but the army naturally suspected the air force. The Democrats in Congress, however, were taking the warnings seriously.

Nineteen fifty-six was, after all, an election year, and any chance to attack Eisenhower for being soft on national security was not to be passed up. Despite his soaring approval ratings, which hovered around 70 percent, the president was vulnerable. He had suffered a heart attack while on vacation in Colorado the previous September, two weeks after the missile meeting in which Nixon had been overruled, and the administration had plunged into turmoil during his lengthy convalescence. Ike's fiercely protective gatekeepers, led by his chief of staff, Sherman Adams (the "Abominable No-Go Man," as he was referred to by jealous colleagues around the White House), restricted all access to the recovering leader. Nixon was not permitted to see him and, in a further blow to his already precarious position, was completely bypassed in the temporary succession of presidential duties. John Foster Dulles, it was decided, would speak for the administration during Eisenhower's incapacitation; the vice president was judged too unseasoned for such an important role. When Nixon did finally see the president, it was only to be told that he should consider taking a lesser cabinet position to get executive experience.

If Nixon chafed at the demotion, he did so gracefully and privately, maintaining a publicly supportive face that earned praise from Ike's inner circle. Eisenhower, though, continued to have misgivings, not only about the competence of his inexperienced vice president but also about his own ability to withstand a grueling reelection campaign and a second four-year term. For four months Ike debated whether to run again, and his memoirs and diaries are filled with the wrenching anguish of that difficult decision. He had never sought high office in the first place and was one of the few American leaders who had never truly aspired to be president. Twice, he had turned down the Republican Party's entreaties to enter the political fray, and in 1952 a write-in campaign had been started without his participation. Sherman Adams, then the governor of New Hampshire, entered Ike's name in his state's primary, and New York governor Thomas Dewey started an Eisenhower-for-president nomination campaign without the candidate. In the end, duty, honor, and a sense of obligation had finally persuaded the war hero to run in 1952, and he had done so with a reluctance that voters perceived as genuine patriotism. Ike's charm was that he was not a career politician but a professional soldier conscripted to serve his country one more time. And now, felled by illness and the creep of old age, when he would much rather retire to his beloved farm in Gettysburg and play golf, he was being asked to do it again. On February 27, 1956—the very day Korolev pitched Khrushchev the idea of using the new ICBM to launch a satellite—Eisenhower, with the utmost reluctance, publicly announced his intention to seek a second term. He did not name Nixon as his running mate, however, telling reporters, "I will never answer another question on this subject until after August," the date of the Republican National Convention in San Francisco.

The Democrats seized on this uncertainty and ambivalence, this chink in the president's otherwise formidable political armor. "Every piece of scientific evidence that we have indicates that a Republican victory would mean that Richard Nixon would probably be President within the next four years," said Adlai Stevenson, the Democratic challenger, in an effort to scare voters.

National security was another way of chipping away at Ike's popularity. After all, Eisenhower had ended the twenty-year Republican White House drought in 1952 by lambasting Truman's weak defense record and promising to keep the country safe from the perfidious threat of global communism. Now he could be hoisted by his own petard. With the coming presidential polls firmly in mind, Democratic senators duly convened "Air Power" hearings on April 16, 1956, to look into the politically promising prospect of what journalists had dubbed the "bomber gap."

Senator Stuart Symington of Missouri gaveled open the hearings. Unlike Eisenhower, Symington did harbor presidential aspirations. Tall, and handsome ("an ex-playboy," *Time* informed its readers), the freshman senator had first garnered national attention in 1954 by taking on Senator Joseph McCarthy, the rabble-rousing Red-baiter whose communist witch hunts had ruined thousands of lives and pitched the United States into a frenzy of right-wing paranoia. "You said something about being afraid," Symington declared, staring down McCarthy during televised hearings. "Let me tell you, Senator, that I'm not afraid of you. I will meet you anytime, anywhere."

The showdown had marked the fifty-three-year-old Symington as a rising star within the Democratic ranks, perhaps too new on the political scene to win the party's 1956 nomination but a serious contender for 1960. "He is a formidable-looking figure," *Time* noted approvingly, "sprawling in his red leather chair, a spectacular executive when transacting business over the telephone. He is abrasive with foot-dragging underlings. He incessantly chomps gum."

Symington, a Yale graduate and the wealthy scion of a patrician political dynasty from New Hampshire, had served as the first secretary of the air force under Truman. For him, the Air Power hearings were not merely a cynical vehicle for self-promotion. He had a genuine and sentimental attachment to the service he had helped found. "We feel, with deep conviction, that the destiny of the United States rests on the continued development of our Air Force," he declared. "The question of whether we shall have adequate American air power may be, in short, the question of survival."

Thus prepped, the senators then heard from a parade of air force and intelligence officials who each offered flimsy but frightening testimony about Soviet heavy bomber production forecasts. By late 1958, they warned, the USSR would have four hundred Bisons and three hundred Bears capable of striking the American heartland. The armada could disrupt the balance of power and lead to a situation where the Soviet Union could actually overtake the United States in intercontinental bomber capabilities. General LeMay, the star witness, reminded congressional leaders that manned strategic bombers were still the weapon of choice: "We believe that in the future the situation will remain the same as it has in the past, and that is that a bomber force well-equipped, determined, well-trained, will penetrate any defense system that can be devised."

As proof of the alleged Soviet buildup, two pieces of evidence were presented. The first was a grainy photograph of the serial number stenciled on the fuselage of a Bison that had flown at a May Day parade in Moscow. It revealed a high numeric series, which implied a vast production line. The second proof was an eyewitness account of U.S. Air Force officers observing squadron after squadron of Bisons doing flybys at the Aviation Day air show in Moscow.

Eisenhower was skeptical, and when Charlie Wilson testified that the intelligence was "very sketchy indeed," Symington indignantly accused him of "unconstitutionally contradicting patriots" like General Nathan Twining, the air force chief of staff. In reality both the flybys and serial numbers were Soviet ruses to mask the weakness of their bomber program. The Soviets simply used the same squadron of planes to circle the airfield out of eyesight and to pass over the reviewing stands repeatedly. Knowing that American observers would have their telephoto lenses trained on the planes, they fudged the serial numbers to further the impression of an inflated count.

The ploy, however, backfired and played right into the hands of the air force and its supporters, who saw a perfect excuse to bolster America's bomber fleet. The truth of the matter was that by the forecast date, the USSR would build only 85 of the 700 new bombers projected by air force intelligence, while the U.S. heavy bomber force would grow to 1,769 planes—a twenty-to-one ratio in America's favor that hardly called for additional reinforcements. (The Strategic Air Command would add another 1,000 bombers to that already overwhelming superiority by the end of LeMay's reign.) But in the absence of hard evidence to the contrary, the alarmist air force assertions were accepted. The accuracy of the hyped data was not pertinent, but its potential political value was. The dearth of reliable information on the Soviet Union simply heightened paranoia and made the worst-case scenario easier to swallow. "You'll never get court-martialed for saying [the Soviets] have a new type of weapon and it turns out that they don't," Victor Marchetti, the CIA's top Soviet military analyst, ruefully remarked. "But you'll lose your ass if you say that they don't have it and it turns out that they do."

The bomber gap was "fiction," as Eisenhower well knew. But the president did not challenge Symington's findings. In fact, many of the air force officers who provided the testimony and information for the hearings were promoted, including the air force's intelligence chief, Major General John Samford, who was rewarded with the top slot at the newly formed National Security Agency. Nor did Ike veto the supplemental $928.5 million budget increase for LeMay to add six more SAC wings—180 new B-52s—to his armada. Boeing, the principal financial beneficiary of the supplement, immediately started a second production line to fill the order.

The billion-dollar boondoggle was the price Eisenhower paid to prevent the Democrats from making national security an election issue. He could not appear dovish, especially since his failing health had left him exposed to criticism. Already the New York papers were hinting that the heart attack and Eisenhower's subsequent stomach surgery for ileitis in early 1956 had debilitated him. Arthur Krock of the *New York Times* acidly speculated whether Eisenhower's "frequent changes of scene and recreation imply that he is irked by his heavy and incessant duties." The president's penchant for delegation, disdain for detail, and notoriously tangled speaking style ("in which numbers and genders collide, participles hang helplessly and syntax is lost forever," according to Krock) offered more grist for allegations of mental torpor.

The growing disenchantment of the press had not yet filtered down to the average voter, who still liked Ike. But, as the historian Fred Greenstein noted, "the much publicized golfing trips, the working vacations, and even the Wild West stories he read at bedtime, which many critics suggested were the outward signs of a passive president with a flaccid mind," left Eisenhower particularly sensitive to accusations of weakness. And so defense was the one area where his administration had to maintain a strong public posture at all cost. The air force got its superfluous bombers. Money would simply have to be siphoned from less politically essential military programs.

3

TRIALS AND ERRORS

The Pandora's box that Nikita Khrushchev had pried open during his secret speech at the Twentieth Party Congress would explode in his face eight days later.

On the morning of March 5, 1956, the continuous blaring of car horns pierced the crisp mountain air over the ancient Georgian capital of Tbilisi. Sirens echoed through the steep cobblestone streets, bouncing off the fifth-century facades of ancient buildings, and the sound rippled down through the valley's orchards, ravines, and steaming sulfur springs below. Near the Palace of Labor, a new and unappealing Soviet structure, a crowd of about 150 people marched down the middle of the road. Their heads were uncovered in a show of respect and bereavement, and they carried red wreaths and large portraits of Joseph Stalin with the corners draped in black crape. It was the third anniversary of the Great Leader's death, an event previously marked by solemn processions throughout the Soviet Union.

But on that day only the Georgians commemorated the passing of their native son. In the rest of the empire there was an official and insulting silence: no ceremonies, no tributes, no mass rallies—just persistent and disturbing rumors that Stalin had been discredited as a brutal tyrant. In Tbilisi, the puzzlement turned to indignation. The Georgians were a fiercely proud people, one of the first nations on earth to have adopted Christianity back in AD 337. Tiny as their mountain enclave was, two of their own, Stalin and Beria, had ruled the endless and chaotic landmass of Russia and had presided over the largest expansion of the Slavic empire since Catherine the Great. Under Stalin, Georgian had become the unofficial second language of a global superpower, and ambitious apparatchiks in Moscow affected soft, slurry Georgian accents. Now, unsettling reports were circulating that Stalin's memory had been besmirched, that the "Great Son of the Georgian People" had been denigrated. In Tbilisi, this caused grave patriotic concern.

The following day, something strange happened. The mourners returned for a spontaneous, unsanctioned march. This time there were one thousand people, and they carried portraits of Lenin in addition to Stalin. The mood was different as well, recalled Sergei Stanikov, the local correspondent for the Moscow daily *Trud*, or Labor. Stanikov, a loyal party man, smelled a story. In twenty years of covering Georgian politics, he had never witnessed a spontaneous rally. No one demonstrated in the Soviet Union without permission. Stanikov followed the crowd as it squeezed through the narrow, musty streets and made its way to Georgian Central Committee headquarters at Government House. Was it true, the mourners demanded, what was being said about Stalin in Moscow? "A meeting was held at 4 o'clock in which I was present," Stanikov wrote in a secret report. "Comrade Mzhavanadze [the first secretary of the Georgian Communist Party] informed us that he would soon acquaint us with the letter regarding the Cult of Personality."

Unsatisfied with this vague and discouraging response, the mourners returned in even larger numbers the next day. At noon, students from Stalin University, the Institute for Agriculture, and the Polytechnical Institute walked out of their classrooms and joined the growing mob on Rustaveli Street. "*Dideba did Stalins*," the crowd chanted defiantly in Georgian, "Long live Stalin." The students were angry. Georgians had a reputation for their fiery temperament; their hospitality knew no bounds, as evidenced by their elaborate twenty-one-toast protocols during brandy-filled banquets for foreign guests, and neither did their anger at perceived slights. A car was overturned. Someone threw a rock at the City Council Building. The chants also took a more ominous, nationalistic turn. "Long live Georgia, long life to the Georgian people."

Local officials dispatched frantic messages to Moscow. They had no experience with civil disobedience and wanted instructions from the Kremlin. This, in itself, was not unusual. In a vertical hierarchy like the USSR, no official who valued his perks—or position—made independent decisions without consulting higher authority. The members of the Presidium were so bogged down with the minutiae of running this vast nation that they were currently reading the manuscript of a young writer named Boris Pasternak to decide whether his novel *Doctor Zhivago* should be banned. (It was.) The Ministry of the Interior told the Georgian officials to publicly read Khrushchev's proclamation on Stalin's crimes and the Cult of Personality. That should silence the crowd. But it didn't.

As Molotov and Kaganovich had feared, Khrushchev's precedent of openly criticizing a Communist icon served only to whip Georgians into their own frenzy. It was as if Khrushchev, by his own example, had given citizens license to criticize the regime as well. And now that they had tasted a little freedom to protest, they couldn't stop themselves.

The following day, March 8, some eighteen thousand demonstrators filled Lenin Square in Tbilisi, their grievances having grown more wide-ranging. "Provocative speeches of inflammatory, chauvinistic and anti-Soviet nature were read," Stanikov reported indignantly to his editors, who relayed the breathless dispatches to KGB headquarters. (Stanikov, like other Russian correspondents in Tbilisi, understood that his field reports were not for publication, and the Soviet media made no official mention of the brewing unrest. Journalists simply acted as an extra set of eyes and ears on the ground for the security organs.) "A huge young man with a Tarzan hairdo," Stanikov continued, "waved his fist in the air, and after a series of accusations against the Party and the government, went on to recall the struggle of the Georgians against foreigners."

Stanikov was shocked, and for the first time since the marches began he was also frightened. The demonstration had shifted to dangerously seditious grounds, because the "foreigners" in question were Russians. What had started as indignation over Khrushchev's denunciation of a hometown hero was rapidly devolving into an independence rally, a cry for autonomy from the Kremlin itself. Stalin, after all, had been the central linchpin to Georgia's tenuous allegiance to Moscow. Without the reflected glory of the Great Leader, it did not take long for Georgians to remember that they had been forced at bayonet point to join the empire.

Poets and writers stoked the nationalist flames with fiery readings of Georgian folklore. People waved purple prerevolutionary Georgian flags and sang odes to kings from centuries past. Every vehicle in the city honked its horn in solidarity. Even Stanikov unwittingly caught the liberty bug, feeling free to criticize Soviet authorities in one dispatch. "In my opinion, the public reading of [Khrushchev's] letter on the cult of personality should have been avoided," he complained to his bosses, expressing a contrarian view to the party line that, in Stalin's day, only a fool would have put in writing.

The situation was spiraling out of control. In Moscow, on the evening of March 8, the Presidium debated what to do. Khrushchev urged restraint; Molotov and Kaganovich wanted order restored immediately. The unrest had started to spread to other Georgian cities. Left unchecked, the madness could infect neighboring regions in the volatile Caucasus and possibly contaminate the entire Union of Soviet Socialist Republics, inflaming the nationalist aspirations of Kazaks, Uzbeks, Tajiks, Tatars, Balts, Moldovans, and western Ukrainians—to say nothing of the Poles, Czechs, Bulgarians, Romanians, Hungarians, and East Germans living unhappily under Moscow's yoke. The Presidium argued late into the night. And by morning the reports from Tbilisi were increasingly alarmist. "On March 9th unimaginable things were happening," Stanikov wrote. "Not only the youth, but even adults were going berserk on the streets. Most of the small workshops were closed. The employees of small offices came out on the streets. . . . The movements of trams, buses and trolley buses were disrupted."

The entire city was paralyzed. The sprawling Stalin Coach Works, the huge locomotive factory, shut down. In the early afternoon, the Georgian first party secretary tried to address the agitated crowd outside Government House but was shouted down. No one had ever publicly defied the senior Communist Party representative before. Terrified, he locked himself in his office, surrounded by armed guards. The Georgian Central Committee hotline to the Kremlin burned with urgent pleas for guidance. The local authorities were no longer in control.

Just before midnight, word reached Moscow that a huge crowd had descended on the main radio and telegraph station in central Tbilisi, demanding to broadcast from the transmission tower. Tanks had deployed around the communications center and were anxiously awaiting instructions. The time for debate was over. Something had to be done now. "This is what the provocation, that was apparently organized by foreign spies and agents, and which was not dealt with in time, led to," Stanikov reported afterward. "Hooligans put everything into action: knives, stones, belts. There was no way out for the soldiers. Their life was in danger and they were forced into taking defensive action."

By the time the T-55 tanks finished "defending" themselves from the onslaught of civilian belt buckles and pocketknives, nine protesters were officially pronounced dead, dozens lay wounded, and thirty-eight alleged ringleaders were arrested. Historians would later increase the body count ten- to twentyfold, though to this day no one knows how many Georgians actually died during the March 9 massacre. One thing was clear, however: Nikita Khrushchev had unleashed powerful pent-up forces with his secret speech. His attempt to breathe a little democracy into the Stalinist corpse had horribly backfired because a little democracy can be dangerous in a totalitarian society. Liberalization is a slippery slope. And dictatorships can easily lose their footing once they loosen the reins. Khrushchev didn't understand that, and he underestimated the longing for self-determination of the nationalities held captive behind the Iron Curtain.

His troubles, in fact, were just beginning.

• • •

As the winter wore on and the KGB hunted for the "nest of foreign spies" that had incited the Georgian uprising, Sergei Korolev grappled with his own share of difficulties. His problems, like Khrushchev's, had been mostly of his own making.

Korolev had fallen into the classic salesman's trap. During his pitch to the Presidium at OKB-1, he had made wildly unrealistic promises in order to close the deal for the satellite. By showing the Presidium a full-scale mock-up of the R-7, he had left Khrushchev and Molotov with the distinct impression that a prototype of the ICBM was almost complete. But the rocket they had seen was an illusion, little more than a ten-story-tall modeler's toy. The real prototype was nowhere near ready. The satellite was also hopelessly behind schedule. And the modifications required for the rocket to carry it were not nearly as "minor" as Korolev had breezily suggested. In short, Korolev had conned Khrushchev, a sucker for engineering marvels who could easily "be beguiled by a charismatic scientist promising miracles," according to his biographer William Taubman.

Korolev had played on Khrushchev's intellectual aspirations and educational shortcomings, while downplaying his own limitations. During the Presidium presentation, he had even silenced Glushko, the suave main engine designer, who had emphasized the daunting complexities of making the R-7 operational. "Our guests are not interested in the technical details," Korolev had interrupted, with a cheerfully dismissive wave. But now that Korolev had to deliver on his promise, those technical challenges were mounting.

Glushko's engines were the first hurdle. They had to be almost ten times as powerful as anything ever built before, and required a radically new design. Their success also depended on the ability of Glushko and Korolev to get along, which was no easy task since they had a long and acrimonious history to overcome. Aside from their mutual disdain, these two titans of Soviet rocketry were diametric opposites. Glushko was elegant and regal, with delicate, slightly feminine features, and soft, sensuous Asiatic eyes that hinted at a genetic link to Mongol invaders from centuries past. Matinee-idol handsome, he took great pride in his appearance. His suits were handmade of imported black-market fabrics. His shirts were cut and starched in the latest Western style. And he wouldn't be caught dead in the Bulgarian and Polish shoes favored by party high-ups. He fussed over his hair and neatly manicured nails, and selected his silk ties carefully.

Korolev, on the other hand, never wore a tie unless he had to, favoring black leather jackets, and he looked like a heavyweight boxing coach who had taken a few punches on the chin. His thick fingers were nicotine-tinged, and his shirts were wrinkled and often stained with soup. His thinning black mane had a will of its own whenever he forgot to slick it down, which was frequently, and it could safely be said that he didn't give much thought to his appearance.

Glushko loved the ballet and classical music, and he enjoyed long, languid meals at Moscow's few fine restaurants. Korolev had no interests outside of rockets and viewed food as fuel. "He ate very quickly," a fellow OKB-1 engineer recalled. "After finishing the food on his plate, he would wipe it clean with a piece of bread, which he subsequently put in his mouth. He even scooped up the crumbs and ate them. Then he licked all his fingers. The people around him looked on with amazement until someone volunteered that this was a habit he had developed during his years in prison and in labor camps."

No one who had ever gone through the gulag emerged physically or psychologically unscathed. Though the Chief Designer rarely spoke of it, the Great Terror had left indelible marks on both his body and his soul. Even decades later, he could remember the minutest details of his arrest in 1938: the rasp of the needle on the gramophone that kept churning its spent record while the men in black ransacked his apartment; the sound of the trolleybus bells ringing six stories below; the hushed whimper of his three-year-old daughter, Natalia, as she clung to her terrified mother.

At the Kolyma mines, the most notorious of Stalin's Siberian death factories, his days had started at 4:00 AM, in sixty-below-zero darkness that lasted most of the year, conditions that killed a third of the inmates each year. The criminals administered justice in Kolyma, beating the political prisoners mercilessly if they dropped a pickax or spilled a wheelbarrow or missed their quota. They stole their food and clothes, and pried out the gold fillings the greedy guards had overlooked.

Within a few months of his arrival at Kolyma, Korolev was unrecognizable. He could barely walk or talk; toothless, his jaw was broken, scars ran down his shaven head, and his legs were swollen and grotesquely blue. Scurvy, malnutrition, and frostbite had started their lethal assault and he seemed destined to die.

And yet, like millions of other purge victims, he still held out the hope that Stalin himself would realize that a terrible mistake had been made and free him. "Glushko gave testimonies about my alleged membership of anti-Soviet organizations," Korolev wrote Stalin in mid-1940, naming several others he claimed had borne false witness against him. "This is a despicable lie. . . . Without examining my case properly, the military board sentenced me to ten years. . . . My personal circumstances are so despicable and dreadful that I have been forced to ask for your help."

Korolev's mother also petitioned Stalin directly. "For the sake of my sole son, a young talented rocket expert and pilot, I beg you to resume the investigations."

But their letters, like the countless other pleas of assistance with Siberian postmarks that reached the Kremlin daily, went unanswered. What saved Korolev, ultimately, was Hitler's 1941 invasion of the Soviet Union and the sudden drastic need for skilled military engineers. Transferred to one of the special Sharaga minimum-security technical prison institutes that Beria had set up to exploit jailed brainpower, Korolev slowly recovered, until, at war's end, he was released and sent to Germany to parse the secrets of the V-2.

Glushko had also been in the camps, and Korolev had worked for him when they were sent to the same Sharaga. He too had been told, or had somehow come under the impression, that Korolev had denounced him during the purges. After all, everyone talked and named names after a few days with the NKVD, the predecessor to the KGB. Korolev himself had signed a written confession of guilt, following one of the bloodier sessions of his interrogation. And so perhaps it was true, even likely, that he had implicated others under duress. Neither man would ever truly know, though each would harbor his suspicions. Such was the fate of the children of a revolution that ate its own and spread complicity like a soul-sapping disease.

In America, the blacklists that had cost scientists and Hollywood writers their careers or promotions during McCarthy's rampage had led to lifelong grudges. Here the betrayals had cost people their freedom and lives, their families and possessions. And yet, when it was all over, the denouncers and denouncees were all thrown back together to coexist peacefully as if nothing had happened. No Westerner could ever hope to understand this peculiarly Soviet condition, this enforced amnesia.

Had Glushko and Korolev forgotten, or forgiven each other? Or had they simply buried their simmering recriminations and resentments beneath a thin veneer of civility?

Another factor complicating their reconciliation was the persistent rumor that Korolev had engaged in a long-running affair with Glushko's sister-in-law. The alleged romance—a source of contention among contemporary Russian historians—had predated the purges and apparently resumed in 1949, as the Chief Designer's first marriage was falling apart. Whether Glushko knew about it, or how he felt about his brother's deception and humiliation at the hands of his rival, is not a matter of the historical record. The only thing clear amid the unanswered questions was that Korolev and Glushko needed each other and had to find a way to work together.

Korolev now outranked Glushko, which probably didn't help heal the old wounds, for the one characteristic the two men shared was pride. Both were intensely egotistical and ambitious and fiercely competitive by nature. In manner and demeanor, however, they were different in almost every other respect. Korolev was coarse and crude, capable of violently profane tirades during which he would scream, shout, and dismiss employees with threats of extended sojourns to Siberia. Boris Chertok recounted witnessing one such outburst. It occurred in 1945, when the two were in Germany, scouring for V-2 technology. As war booty Korolev had procured a shiny red Horche two-seater sports car, which he drove at breakneck speed, terrorizing passengers and pedestrians alike. "Sergei Pavlovich," Chertok had pleaded, "your Horche is beautiful, but it's not a fighter plane, and we are in a populated area, not the sky."

"But I have both a driver's license and a pilot's license," Korolev had retorted confidently. Sure enough, a few days later Korolev rammed a vehicle from the Soviet carpool just outside Chertok's headquarters. "Korolev flew into my office extremely upset and demanded I immediately fire the German driver, and send Chiznikov [the Russian officer in charge of transportation] into exile for not keeping order in his motor pool."

Korolev's anger, however, never roiled for long. His temper subsided as quickly as it flared, and he would return a few hours later with sheepish good cheer, as if nothing had happened. (Chiznikov, for instance, was never banished for vehicular disorderliness; instead, Korolev promoted him and recruited him to work at OKB-1, where he liked to boast to impressionable young engineers: "I'm not afraid of anyone in the whole wide world—except Korolev.")

Glushko, conversely, never betrayed emotion or raised his voice in anger. But he held bitter and enduring grudges and worked mercilessly behind the scenes to exact revenge on those who crossed him. He was a formidable adversary, and like Korolev he knew how to work the system. Neither man was much for compromise. And theirs was an uneasy partnership destined to explode.

The first hurdle the pair faced in the wake of the Presidium visit was rocket power. To fling a five-ton thermonuclear warhead 5,000 miles, Korolev had calculated that he needed more than 450 tons of thrust. This represented a tenfold increase over the R-5 intermediate-range ballistic missile, whose RD-103 booster had already strained the limits of single-engine capacity. (Glushko had experimented with a mammoth 120-ton thrust engine, the RD-110, but the project had been abandoned.) To meet Korolev's vastly increased power requirement, the R-7 would have to be powered by five separate rockets, bundled like a giant Chinese fireworks display. But that still left Glushko with combustion chamber problems. Even with five supercharged furnaces sharing the load, they would still be too big and unwieldy, subject to destabilizing power fluctuations. The solution was the RD-107, an engine with a single turbo pump that simultaneously fed fuel and oxidizer to four regular-sized combustion chambers. The combined output of the twenty combustion chambers—four in each of the five boosters— would meet Korolev's thrust specifications. The clustering of the peripheral boosters around the central rocket would make the R-7 ungainly in appearance with a bulging thirty-five-foot-wide base that resembled the skirt of a hefty babushka, but the RD-107s were far more efficient than previous-generation motors. With recast combustion chambers, shaped like cylinders rather than the standard flared mushroom mold popularized by the V-2, they would ignite kerosene, which burned hotter than the alcohol-based propellants the Germans and Americans used.

In addition to generating more lift, the five-booster configuration solved another serious design concern. The Soviets had never designed a multistage rocket before, and Glushko refused to guarantee that the upper stages—so critical to achieving orbital velocity—would work. But if the central R-7 engine block was designed to operate longer, the four peripheral boosters could be jettisoned in flight to lighten the load, while the central core kept firing in what was effectively a one-and-a-half-stage compromise.

The solution, however, created its own new snag. Because the larger central engine now worked longer, about four minutes in total, the heat-resistant graphite steering vanes Glushko had fixed to the exhaust nozzles could not be employed. They were rated to withstand the intense heat for only two minutes before burning up, and without them the R-7 could not be steered. A pitched debate broke out over how to fix the problem. Korolev favored using small gimbaled thrusters, mini-combustion chambers on swivels to provide steerage. But Glushko was violently opposed to the idea of anyone tinkering with his engines. (This was a man who so hated outside interference that he once famously drove several hundred miles with the handbrake engaged rather than heed a passenger's advice to release it.) Already with the R-7, he was making a significant concession by using kerosene and liquid oxygen propellant rather than his favored mix of nitric acid and dimethyl hydrazine. And now Korolev was demanding even more significant alterations. A shouting match ensued, with Glushko hotly refusing to budge. If the Chief Designer wanted the damn changes, he'd have to make them himself.

Friction was also posing serious difficulties at the working end of the R-7. At the speed the missile would travel, none of the nose cones in Korolev's inventory could withstand the heat that would be generated during the 24,000-feet-per-second atmospheric reentry. And without that thermal protection, the warheads would be incinerated, rendering the ICBM useless.

The problems mounted. Tests showed that the standard support blocks used to prop up rockets at launch would not support the mammoth R-7, which at 283 tons weighed more than any object that had ever been flown. A huge gantry rig with the rough dimensions of the Eiffel Tower had to be specifically built. So did an enhanced new guidance system, because the almost tenfold increase in the R-7's range magnified minute trajectory inaccuracies by hundreds of miles at the point of impact. And then there was the nagging issue of postimpulse boost, the unwanted thrust from fuel remaining in a rocket's plumbing system after the engine shut down. With small rockets, extraneous fuel amounted to only a few gallons that burned off harmlessly without affecting trajectory. But with a missile the size of the R-7, the amount of propellant left over in the feed lines was significantly larger. The residue could keep the engines firing for as much as a full second after shutdown and push the missile hopelessly off course at the critical aiming point.

All these were normal glitches, typical in the creation of any major new weapons system. Korolev, however, had promised Khrushchev an impossibly tight schedule: to have the R-7 ready for flight testing in January 1957. Khrushchev, in turn, had started chopping military spending for conventional forces in anticipation of the ICBM becoming Russia's main line of defense. He had halted the construction of costly aircraft carriers and heavy cruisers, infuriating his admirals, and had cut back on the production of long-range bombers, outraging his air marshals. If Korolev didn't deliver, he could be jeopardizing the security of the Soviet Union, to say nothing of risking the Presidium's collective wrath. Khrushchev may not have been a butcher like Beria, but the men around him were, and they would have no hesitation making their displeasure known.

The Chief Designer began to sweat. He could not have forgotten the telephone call he had received from Beria in 1948 after the R-1 had suffered several setbacks. "I've been sent the protocol of the latest tests," Beria said, his childlike voice barely a whisper. "Another failure. And again no one is to blame. Some people are soliciting an award for you—but I think you deserve a warrant!"

Korolev had frozen in fear. "We are doing our work honestly," he stammered. But Beria had no interest in excuses. "Do you understand about the warrant?" he hissed, hanging up.

Already the R-7's delays were piling up, and now Glushko had seriously bad news. Static firing tests of his new engines showed the boosters were not performing to expectations. Thrust was not the issue. In terms of brute force, Glushko's motors had generated 396.9 tons of lift, which translated to a better than expected 490.8 tons in the vacuum of space. The trouble was in the all-important realm of specific impulse, the ratio that calculated how many pounds of thrust were produced per each pound of propellant consumed per second. It was the aeronautic equivalent of the automobile industry's gas-mileage ratings, which determined the fuel efficiency of different cars. The specifications for the R-7 had called for a specific impulse of 243 at sea level and 309.4 in space. But Glushko's engines had come up short, at 239 and 303.1 respectively, a serious enough setback to warrant an official communiqué to the Kremlin. "At present time, we are completing static testing of the rocket," Korolev glumly informed the Central Committee in the fall of 1956. "Preparations for the first launch of the rocket are experiencing substantial difficulties and are behind schedule. The current results of the stand tests give us solid hope that by March of 1957 the rocket will be launched. After small modifications, the rocket can be used to launch an artificial satellite with a small payload of scientific instruments of about 25kg."

Couched in the soothing verbiage of "solid hope," Korolev was effectively warning Khrushchev that not only was he going to miss his January deadline, but the R-7 probably wouldn't be ready in March either. What's more, instead of the 2,200-pound advanced satellite package he had promised, the R-7 would manage to carry only a meager fifty-pound payload into space because of its poor fuel efficiency.

The reality check, Korolev well knew, was not likely to sit well with the impatient Presidium. Fortunately for the Chief Designer, Khrushchev and the Central Committee were preoccupied with other, far more urgent matters.

• • •

On the morning of June 28, 1956, workers in the western Polish city of Poznan declared a general strike. It was the first labor unrest in the Soviet bloc, and by early afternoon the walkout had turned into the largest anti-Communist rally since the war. One hundred thousand people, a third of Poznan's population, crammed Adam Mickiewicz Square, waving banners that read DOWN WITH DICTATORSHIP and, in a play on Lenin's most famous revolutionary slogan, WE WANT BREAD, FREEDOM, AND TRUTH.

As Kaganovich and Molotov had feared, the ill winds of liberalization let loose by Khrushchev's de-Stalinization decree had blown westward from the snowcapped Caucasus to the plains of Poland. Only Poland wasn't an isolated and inaccessible mountain republic with a few million inhabitants. It was smack in the middle of Europe, the anchor of the new Warsaw Pact defense league designed to counter the North Atlantic Treaty Organization, and by far the largest of the Kremlin's foreign holdings. It was also a very reluctant Soviet vassal, a nation with a powerful peasant class that refused to collectivize, an activist Catholic Church that could not be brought to heel, and a thriving black-market economy that was capitalist in every respect but name. It had been the unruly Poles who had leaked the contents of the secret speech to Israeli intelligence and the CIA, and it was Poland, more than any other Soviet satellite state, that had taken Khrushchev's reformist message to heart. The thaw in Poland had actually begun shortly after Stalin's death, but Khrushchev's encouraging speech had accelerated the process, giving Poles new hope for democratic freedoms. Censorship in the media melted away, people grumbled publicly, and dissident officials who had been imprisoned or expelled from the Polish Communist Party were officially rehabilitated.

In Poznan, trouble had been brewing for weeks. Residents were unhappy with working conditions, housing and food shortages, and low pay. When they marched in protest, the crowd's anger quickly found an outlet in the secret police headquarters on Kochanowski Street. Tens of thousands of people, some armed with bats, pipes, or cobblestones ripped from the road, laid siege to the security service building. A shot rang out from one of the upper windows, and thirteen-year-old Romek Strzalkowski fell dead. After that, things got ugly. Marshal Konstantin Rokossovsky, the Soviet Union's military overseer for Poland, ordered tanks and ten thousand troops into Poznan. This time, the official body count was seventy dead, three hundred wounded, and seven hundred arrested.

By October, Rokossovsky was massing troops outside Warsaw and the Soviet embassy was sending distress signals to Moscow. "The Poles were vilifying the Soviet Union, and were all but preparing a coup to put people with anti-Soviet tendencies in power," Khrushchev wrote in his memoirs. The Central Committee of the Polish Communist Party was now itself in open revolt. Poland's Communist leaders wanted to oust their Moscow-backed first secretary and replace him with Władysław Gomułka, whom Stalin had jailed for "nationalist deviation." Khrushchev was shocked. His rehabilitation program never envisioned actually putting former political prisoners into positions of high authority, never mind as heads of state. But the residents of Warsaw were arming themselves, preparing to repel Rokossovsky's troops. On the morning of October 19, Khrushchev, Bulganin, and a visibly agitated Molotov flew to Warsaw. "From the airport we went to the [Polish] Central Committtee," Khrushchev recalled. "The discussion was stormy. . . . I saw Gomułka coming toward me. He said, very nervously, 'Comrade Khrushchev, your tank division is moving toward Warsaw. I ask that you order it to stop. I'm afraid that something irreparable could happen.'"

Khrushchev halted the tanks and accepted Gomułka's leadership. "A clash would have been good for no one but our enemies," he reasoned. Faced with the imminent prospect of armed intervention, Gomułka, in turn, promised to keep Poland in the Soviet orbit. They had gone to the brink, but catastrophe had been narrowly avoided. Two weeks later, under almost identical circumstances, no one would be so lucky.

• • •

South of Poland, in Hungary, a crisis had been building since October 23, when students in Budapest began demonstrating in solidarity with Gomułka's defiance. Quickly the protests spread, as factory and office workers joined the revolt. The crowds marched on the main radio tower to broadcast their grievances, what they called their Sixteen Points: more freedom and food, less police interference, fewer travel restrictions, and the withdrawal of Soviet troops.

Radio Free Europe, the U.S.-sponsored network that beamed Western news into the Eastern bloc, picked up the rallying cry, its broadcasts becoming increasingly insurrectionary. Secretary of State Dulles, who had long vowed to "roll back" communism, encouraged the Hungarian demonstrators and pledged American support. "To all those suffering under communist slavery," he said, "let us say you can count on us."

Emboldened, the protesters surrounded parliament and gathered outside the secret police headquarters, chanting for the Red Star atop the building to be removed. Their requests were met by a hail of gunfire, and in the ensuing hand-to-hand fighting eighty were killed. But the building was taken, as was a nearby armory. The uprising was now an armed revolt, gaining momentum. Radio Free Europe stoked the flames, instructing Hungarians on how to make weapons out of gasoline, bottles, and rags. Hungarians felt certain that the United States was behind them. Within days, 80 percent of the Hungarian army had switched sides. The Hungarian people could taste the liberty that Dulles had promised.

In Moscow, at an October 28 emergency session of the Presidium, Khrushchev counseled caution. Molotov and the hard-liners wanted swift action, but the first secretary advocated compromise. "The Soviet government is prepared to enter into the appropriate negotiations with the government of the Hungarian People's Republic, and other members of the Warsaw Treaty, on the question of the presence of Soviet troops on the territory of Hungary," *Pravda* announced on October 31.

In Washington, a jubilant Allen Dulles hailed the concession as "a miracle," the most meaningful sign yet that communism could be in retreat. "This utterance is one of the most significant to come out of the Soviet Union since World War II," he told Eisenhower.

"Yes," the president agreed skeptically. "If it is honest."

Khrushchev, the reformer, had unwittingly opened the floodgates, and now the Kremlin was being swamped in a tide of upheaval. And if Hungary fell, Moscow's other dominions would quickly follow. No one would be able to stop the outpour.

Just as swiftly, however, the tide turned. In Budapest that same day, the violence spiraled into an orgy of revenge. Dozens of suspected secret police officers and informers were hung from lampposts by rampaging mobs, while Imre Nagy, the Hungarian prime minister, defiantly summoned the Soviet ambassador, Yuri Andropov—the future KGB boss and head of state. Like Gomułka, Nagy had been imprisoned by Stalin and rehabilitated after the secret speech. Under Khrushchev's liberalizations he had replaced the Stalinist puppet Mátyás Rákosi only a few months earlier. And now he had done the unthinkable. Hungary, Nagy told Andropov, was renouncing the Warsaw Pact and proclaiming its neutrality. A telegram had already been dispatched to the United Nations, asking the United States, Britain, and France "to help defend" the breakaway Soviet satellite.

When news of the declaration and lynch mobs reached Moscow, Khrushchev reversed course. "We have no choice," he said at another emergency Presidium session that evening. "We should take the initiative in restoring order."

"Agreed," growled Molotov, Bulganin, and Kaganovich. "We showed patience but things have gone too far. We must act to ensure that victory goes to our side."

"If we depart from Hungary," Khrushchev went on, "it will give a great boost to the imperialists. They will perceive it as weakness on our part and go on the offensive."

"We should use the argument that we will not let socialism in Hungary be strangled," volunteered Pyotr Pospelov, the party's PR chief and the editor of *Pravda*. "That we are responding to an appeal for assistance."

Khrushchev agreed. Closing the session, he instructed Marshal Georgy Zhukov, his deputy defense chief, "to work out a plan and report it."

There was no longer room for negotiation. This time an example had to be made; the fate of the Soviet empire depended on it.

"Bombs, by God!" Eisenhower was awakened in the early hours of November 1. But it was not a Soviet invasion. Instead the news was that France and Britain (along with Israel) had attacked Egypt, in a punitive strike against President Gamal Abdel Nasser's nationalization of the Suez Canal. "What does Anthony think he's doing?" Eisenhower demanded, as he picked up the phone to call the British prime minister, Anthony Eden. The whole world was going up in flames, on the eve of the U.S. presidential election. How could the United States now condemn the Soviets, a furious Allen Dulles lamented, "when our own allies are guilty of exactly similar acts of aggression?"

Now there was no longer any room for talk of coming to Hungary's aid. Moscow would have free rein to teach the Hungarian "hooligans" a lesson no one in Eastern Europe would soon forget. Shortly before 9:00 AM on November 4, the BBC interrupted its regularly scheduled broadcast with the following announcement: "The Soviet Air Force has bombed the Hungarian capital, Budapest, and Russian troops have poured into the city in a massive dawn offensive."

The death toll, this time, was thirty thousand. Russian tanks rolled out of their Hungarian bases and dragged bloodied corpses through Budapest's central squares to serve as brutal warnings to future counterrevolutionaries. Nagy was executed. Thousands were placed under arrest, while two hundred thousand Hungarians fled to the West. By November 14, order had been restored, but there was little doubt in the minds of the Kremlin hard-liners who was truly responsible for the string of rebellions. "Khrushchev's days are numbered," Allen Dulles predicted.

4

TOMORROWLAND

Much as the Anglo-French incursion into the Sinai Peninsula enraged Eisenhower, the simultaneous conflicts in Egypt and Hungary proved a boon at the ballot box. Ike had already enjoyed a significant lead over his Democratic challenger, Adlai Stevenson, when the twin crises erupted on the eve of the presidential poll, but he was ideally positioned to benefit from the international turmoil.

As the incumbent, Ike was able to rise above the political fray and act the statesman, holding emergency meetings of the National Security Council and conferring with world leaders. He called for United Nations resolutions and addressed the nation on television. In those final days of the 1956 campaign, Eisenhower was not running for office; he was brokering peace in the Middle East and trying to contain the carnage in Eastern Europe. Stevenson had little choice but to back the president or risk appearing as if he was putting his own electoral ambitions ahead of the national good.

When the votes were tallied on November 6, Eisenhower had won in one of the largest landslides in U.S. history, carrying nearly 60 percent of the popular vote. Even Nixon's presence on the ticket (the vice president had finally confronted Eisenhower and forced the issue of his renomination) could not dampen the enthusiasm for the president; Ike made significant inroads in the heart of Dixie, giving the Republicans hope that they might one day break the Democratic stranglehold over the South.

With Eisenhower assured of another term in office, the administration no longer had to pander to Symington and the other Air Power hawks. The nearly one-billion-dollar supplemental appropriation for B-52 bombers that the Democrats had pushed through Congress in the run-up to the election had thrown the budget into deficit, and now it was time to balance the books.

Just where the additional B-52 funds would come from became apparent on November 26, when Engine Charlie Wilson announced the Pentagon's new "roles and missions directive." The document was aimed at clearly delineating each of the services' duties and responsibilities, and laying out jurisdictional boundaries for the squabbling chiefs of staff. Bruce Medaris's heart must have sunk when he read the section pertaining to missile development.

> In regard to the Intermediate Range Ballistic Missiles, operational employment of the land-based Intermediate Range Ballistic Missile system will be the sole responsibility of the US Air Force. Operational employment of the ship-based Intermediate Range Ballistic Missile system will be the sole responsibility of the Navy. The US Army will not plan at this time for the operational employment of the Intermediate Range Ballistic Missile or for any other missiles with ranges beyond 200 miles.

The army had been frozen out. Two hundred miles was the range of the existing Redstone missile. The Jupiter, the ABMA project on which the army had rested its hopes, was effectively being turned over to the air force. To add insult to injury, the air force issued a triumphant statement after the ruling, declaring that "it will be better for the country if the ABMA team were broken up and the individuals filtered out into industry and other organizations."

Not only had the air force won the IRBM sweepstakes, now it wanted Wernher von Braun and his German colleagues as well. ABMA might well turn out to be the shortest command in Medaris's colorful career. The irony, he might have reflected, was that Stuart Symington, the man who had set the current fiscal crunch in motion, was the very politician to whom Medaris indirectly owed his first general's star and his belated promotion to brigadier general. It had been during another set of Senate investigations, back in 1953, that Medaris, then a colonel, had learned the power of public relations. Called to testify about the shortages of munitions and equipment in Korea, Medaris brought a grenade to a Senate Armed Services subcommittee, which he dismantled before the startled and spellbound legislators, to expose its precision parts. "If we let down our standards to speed production," he explained, "we do so at the peril of our troops who deserve the safest and most effective firepower we can provide." When confronted with a damning letter from a soldier who had written his mother pleading for her to send him ammunition in her next care package, Medaris, nonplussed, asked to see the document. Flashing a telegenic and slightly roguish smile, he disarmed Symington, who headed the subcommittee, by pointing out that the request was for .32-caliber rounds, which were used for target practice in recreational handguns, and not for military issue. For deflecting congressional criticism away from the army, Medaris within weeks received the promotion that had been denied him for nearly ten years. Politics could give, but it could also take away. Unfortunately, Medaris now ruminated, the air force was proving far more adept at politics than missile development.

The air force, he fumed, was no match for the team of engineers he supervised at ABMA. In fact, it was so short on qualified technical experts that it had been reduced to hiring outside contractors to supervise its existing operations. The Thor program, for instance, was being managed by a recently formed private engineering concern, the Ramo-Woolridge Corporation (later known as TRW). "The lack of a sound, experienced, military-technical organization in the Air Force has been responsible for the technical side of that service becoming almost a slave to the aircraft and associated industries, subject to endless pressure and propaganda," Medaris wrote angrily years later.

Medaris was convinced that his in-house team had more experience and a more proven track record. But the air force had better political connections, along with the deep-pocketed support of the defense industry. So despite the fact that an early prototype of the Jupiter had just flown 3,000 miles, breaking every U.S. record for distance, height, and speed, it was ABMA that was going to be sacrificed to pay for bombers that could be rendered obsolete by the time they rolled off the assembly line. For this, Medaris partly blamed Eisenhower. "In all honesty, I do not think that the situation has been helped by having a soldier in the White House," he complained. "Anyone whose personal experience ended shortly after [World War II] cannot hope to be abreast of today's military needs. Yet having been immensely successful as a theatre commander in a major war, the President is necessarily impressed with his own military knowledge, and less inclined to listen to the advice of today's military professionals."

Eisenhower, however, was more attuned to rocketry's deadly potential than Medaris believed. "Can you picture a war that would be waged with atomic missiles?" the president had asked reporters at a February 8, 1956, press conference. "It would not be war in any recognizable sense." War was a "contest," a battle of wits, strategy, and attrition. Missile warfare, Eisenhower lamented, "would just be complete, indiscriminate devastation." Ike understood the impact of fully developed missiles on any future conflict with the Soviet Union. He was simply in no hurry to rush headlong into what he labeled "race suicide."

The president was far from the only military man with doubts about modern rockets. The air force itself viewed missiles with extreme skepticism, and the bomber generals who dominated the service preferred the proven over the uncertain. In August 1956, for instance, the air force's top research and development commander, General Thomas S. Power, warned of a "somewhat distorted and exaggerated picture" of missile capabilities and complained that missiles "cannot cope with contingencies." And General LeMay openly opposed rockets, which he put at the bottom of his list of military priorities. Missiles, he argued, would gain only a "satisfactory state of reliability" after "long and bitter experience in the field." Meanwhile, they would draw away funds from badly needed bombers. Another general, Clarence S. Irvine, groused that missiles didn't have much of a deterrent effect. "I don't know how to show . . . teeth with a missile," he scoffed.

Nor were these isolated views. Virtually all of the air force's top officers were former bomber commanders or fighter pilots, who saw little glory in sitting in a bunker with a slide rule, pushing launch code sequences. Resistance to missile development within the service was becoming such an issue that Vice Chief of Staff Thomas D. White, in a 1956 speech to the Air War College, warned his subordinates to get with the program.

> We see too few examples of really creative, logical, far-sighted thinking in the Air Force these days. It seems to me that our people are merely trying to find new ways of saying the same old things about air power without considering whether they need changing to meet new situations and without considering the need for new approaches to new problems.

And so the air force was getting missiles it didn't even want, while the army, which desperately wanted them, was being left out in the cold. If Medaris thought the system cynical and Eisenhower out of touch, he was savvy enough to keep his grievances private while he still wore the uniform of a major general. One of his top aides, however, felt no such compunction to suffer in silence. Colonel John C. Nickerson had worked on the Jupiter program since its inception. More than anyone else, he had shepherded it through the labyrinth of the military bureaucracy, had nurtured its various stages of technical evolution, and had rallied behind its creators whenever morale flagged. The Jupiter was his project, and he could not bear to see it aborted. "The aircraft industry, and particularly the Douglas Aircraft Co. [which built B-47 long-range bombers under license from Boeing], openly opposes the development of any missile by a government agency," Nickerson wrote in a lengthy report, which he sent to the aerospace writer Erik Bergaust and the syndicated columnist Drew Pearson. "It is suspected that the Wilson memorandum has been heavily influenced by lobbying by this company, and by the Bell Telephone Co." Bell Laboratories, which had been headed by Donald Quarles for nearly twenty-four years before he became assistant secretary of defense, provided the radio guidance system for the Thor. "Discontinuance of Jupiter," Nickerson went on, "favors commercially the AC Spark Plug Division of General Motors," which was also one of Thor's prime contractors.

This was a direct attack on Secretary Wilson, who had started his GM career at AC Spark Plug and had been the division's president. The accusation implied impropriety at the highest levels of government, and reaction in Washington was swift. Nickerson was arrested, charged with revealing state secrets, court-martialed, and sent to the Panama Canal Zone.

Medaris scrambled desperately to distance himself from the Nickerson debacle, going so far as to testify against him at his trial. But it was too late. The damage had already been done. ABMA had made powerful enemies in Washington, and its future, and that of its star expatriate scientists, looked bleak.

• • •

Not for the first time, Wernher von Braun must have wondered if coming to America had been a mistake. After all, he had chosen the United States, among all the countries who had vied for his services, because he had thought that only America had the resources and foresight to pursue rocket technology. Even before the war had ended, von Braun had gathered his key engineers to discuss which Allied nation offered the best hope for continuing their careers. "It was not a big decision," recalled the physicist Ernst Stuhlinger, one of those present during the defection discussions. "It was very straightforward and immediate. We knew we would not have an enviable fate if the Russians would have captured us." The French had been discounted as strutting losers. The British had fought bravely, but the United Kingdom was small and no longer the power it had once been. West Germany would have strict limits placed on its military programs. That left only the United States.

But the America that greeted von Braun when he first stepped off a military cargo plane in Wilmington, Delaware, in September 1945 was not the place he had expected. At the time, his mere presence on U.S. soil was deemed sufficiently sensitive that it was kept secret for over a year. He cleared no customs and passed through no formal passport controls. The paper trail documenting his entry was sealed in an army vault, along with his incriminating war files; his Nazi Party ties, his depositions denying his involvement in slave labor, and his three SS promotions remained classified until 1984, seven years after his death. For Colonel Ludy Toftoy and von Braun's army minders, his past, as well as the even more frightening war record of some of the other scientists, wasn't an issue. "Screen them for being Nazis?" a senior intelligence officer told the historian Dennis Piszkiewicz, laughing. "What the hell for? Look, [even] if they were Hitler's brothers, it's beside the point.

Their knowledge is valuable for military and possibly national reasons." The real concern in postwar Washington was how the transfer of technology from the Third Reich would play out with the general public. The horrors of the Holocaust were still too fresh, the newsreels of liberated concentration camps still too painful for most Americans to accept Germans in their midst. Toftoy needed to find a secluded spot to stash von Braun and his cohorts until the wounds of the Second World War had healed a little. The hideout he selected was Fort Bliss, a desolate army base near El Paso, Texas. It would become von Braun's less-than-happy home for the next five years.

When Germany's leading scientists began arriving under military escort at Fort Bliss in the fall of 1945, it is not difficult to imagine the culture shock they must have experienced. Brown dusty plains stretched to the east as far as the eye could see. The desert was unbroken save for the occasional tumbleweed, buzzard, and cactus, and it baked at over a hundred degrees for most of the year. To the west rose the jagged red peaks of the Sangre de Cristo, or Christ's Blood, Mountains. To the south ran the Rio Grande and the squalid pueblos of Mexico. At night, an impregnable blackness descended over the land, Stuhlinger recalled. And during the day, the hot Texas sky broiled a deep blue completely alien to any European.

"To my continental eyes," von Braun later confessed, "the sight was overwhelming and grandiose, but at the same time I felt in my heart that I would find it very difficult ever to develop a genuine emotional attachment to such a merciless landscape."

Fort Bliss, an old cavalry outpost built around the rough adobe walls of a border citadel, offered little respite from the inhospitable terrain. Low, ramshackle "rat shack" two-story barracks, made mostly of plywood, sat on the sand. A half-empty hospital building dominated the grounds, along with a few disused hangars and single-story structures connected by gravel pathways. Chain-link fences cut off the German compound from the rest of the base.

Fort Bliss was a far cry from the resplendent accommodations von Braun had grown accustomed to in his German headquarters on the island of Peenemünde, where the sand was confined to pleasant beaches, his sailboat and personal Messerschmitt plane were always at the ready, and the maître d' at the research center's four-star Schwabes Hotel stocked the wine cellar with the finest vintages seized from France. The restaurant's embossed china, fine silverware, and elaborate dining protocols had astounded the Soviets when they occupied Peenemünde. "A line of waiters in black suits, white shirts, and bow-ties marched in solemn procession around the table," Boris Chertok recalled. "In this process, the first waiter ladled soup, the second placed a potato, the third showered the plates with greens, the fourth drizzled on a piquant gravy, and finally the fifth trickled about 30 grams of alcohol into one of the numerous goblets . . . a scene that to us was familiar only from movies." The quality of the ingredients, alas, had deteriorated by the time the Red Army arrived. "I served the best wines," the maître d'hôtel had apologized to his new Communist customers. But when "von Braun evacuated Peenemünde, they took all the food and wine stores with them."

At Fort Bliss, snakes slithered around the cinder-block footings of the mess hall, where cooks in greasy T-shirts slopped something called grub on tin trays. (Complaints about American cuisine figured prominently in the first published press reports revealing Germans in El Paso. The scathing comments of Walther Riedel, von Braun's chief design engineer, earned him a subheading in a December 1946 article: "German Scientist Says American Cooking Tasteless; Dislikes Rubberized Chicken.") The revelation that von Braun's team was in the country in turn prompted impassioned complaints. On December 30, 1946, Albert Einstein and the Federation of American Scientists wrote to President Truman, arguing, "We hold these individuals to be potentially dangerous carriers of racial and religious hatred. Their former eminence as Nazi party members and supporters raises the issue of their fitness to become American citizens and hold key positions in American industrial, scientific and educational institutions." Representative John Dingell, a Democrat from Michigan (whose son holds his congressional seat today), went even further: "I never thought we were so poor mentally in this country that we have to import those Nazi killers."

For von Braun, Einstein's rebuke must have particularly stung. As a teenager in Germany, he had worshiped the theoretical physicist and had written to him, showing his own mathematical equations for spaceflight. Einstein had responded encouragingly, saying that von Braun would make a fine engineer. Now the Nobel laureate was questioning whether von Braun was even fit to be a citizen, much less a scientist.

American citizenship, though, was still far down the pike for von Braun and his German compatriots. As wards of the army, von Braun's group was confined to a six-acre section of the base, kept isolated from other Fort Bliss personnel, and allowed no contact with the local population. "Daily life was quite regulated due to security requirements," recalled Colonel William Winterstein, one of von Braun's early military minders. "The dread that any of the German team may become involved in a public disturbance or accident hung over our heads at all times during those days before it was announced officially that they were in the United States."

Eventually, the restrictions were loosened. In 1947, the thirty-five-year-old von Braun was allowed to return briefly to Germany—accompanied by armed guards—to marry his seventeen-and-a-half-year-old cousin, Maria. At Fort Bliss, security also gradually eased. Once a month, and then once a week, the Germans were bused to El Paso in escorted groups of four to spend a leisurely Saturday afternoon. They were issued IDs that read: "SPECIAL WAR DEPARTMENT EMPLOYEE. In the event that this card is presented off a military reservation to civilian authorities . . . it is requested that this office be notified immediately . . . and the bearer of this card NOT be interrogated." The message was clear; the Germans were the property of the U.S. Army. And only under the supervision of armed MPs could they catch a screening of *Zorro* at the Palace Theatre, go shopping at the Popular Dry Goods Company Department Store, have lunch at the Hotel Cortez, or drink a beer at the Acme Saloon beneath sepia-toned photos of old gunslingers. The outlaw spirit of Wyatt Earp was very much alive in El Paso in the 1940s, and the place still had the rough-and-tumble feel of a frontier town. The streets were dusty and only partly paved. Pickup trucks and the occasional horse-drawn cart plied North Mesa Avenue. Pioneer Plaza swarmed with itinerant farmhands and migrant workers from Mexico. And the Parlor House bar district saw its fair share of fistfights between roughnecks and ranchers, freewheeling businessmen and fire-breathing preachers, who served notice on sinners from tow-away churches in mobile homes.

Wernher von Braun was not a natural fit in this mix. Born to an aristocratic family in 1912, he had been raised on estates in Silesia and East Prussia and in the family residence in Berlin. His father had served as a minister in the government that Hitler toppled. His mother, the Baroness Emmy von Quistorp, was of Swedish noble lineage, spoke six languages, and had been raised in England as a Renaissance woman with interests in astronomy and classical music. She passed her passion for astral and orchestral movements on to her son, who by the age of six had composed his first piano concerto and received his first telescope. In Peenemünde, which von Braun had selected as a research center based on the baroness's recommendation, he pursued his love of music, playing cello in a string quartet of rocket scientists. His group had had its own private chamber in the Officers' Club, Stuhlinger remembered: "His cello was accompanied by Rudolf Hermann's and Heinrich Ramm's violins, and by Gerhard Reisig's viola, when the four of them played works by Mozart, Haydn, and Schubert."

At Fort Bliss, the Officers' Club was off-limits to the Germans. They had to build their own clubhouse in a storage shed, stocking it with homemade furniture and a bar they cobbled together from spare planks. Sometimes von Braun's brother Magnus, who had been a supervisor at Mittelwerk, played the accordion. At other times, after a few rounds of tequila, when the monotony, seclusion, and language lessons got to them, they crept through a hole in the fence to look at the stars in the desert sky. "Prisoners of peace" was how they referred to themselves. But at least they were out of reach of the prosecutors and the war crimes tribunals in Europe that were meting out justice for the slaughter of slave laborers at Mittelwerk, among a host of other Nazi atrocities.

A year, and then two, passed aimlessly. Resources at Fort Bliss were as rare as rain. Only $47 million had been allocated in 1947 for total U.S. missile development, and that left precious little for the Germans. Fort Bliss's miserly quartermaster turned down a request by von Braun's brother for linoleum to cover the cracks between boards in the wood floor of the hut where delicate gyroscopes were assembled. He also denied a requisition for a high-speed drill. "Frankly, we were disappointed," von Braun recalled years later. "At Peenemünde we had been coddled. Here you were counting pennies . . . and everyone wanted military expenditures curtailed."

Von Braun bubbled with ideas for new rockets. But his every proposal was shot down. Nor was anyone particularly interested in his ideas for space travel either. He sent a manuscript on exploring Mars to eighteen publishers in New York, and eighteen rejection letters wended their way back to the Fort Bliss postmaster. Adding insult to injury, von Braun now had to report to a twenty-six-year-old major whose sole technical background was an undergraduate engineering degree. When von Braun was twenty-six, thousands of engineers had answered to him. His loyal Germans still insisted on calling him Herr Professor out of respect. But his pimply new American boss, Major Jim Hamill, addressed him as Wernher and didn't even bother to respond to most of his plaintive memos requesting more materials. Just tinker with your old V-2s was the standing order from Colonel Toftoy in Washington. "We'll put you on ice," Stuhlinger recalled Hamill saying. "We may need you later on."

The U.S. government had gone through a great deal of trouble to ensure that no other power acquired the services of Nazi Germany's rocket elite. Retaining "control of German individuals who might contribute to the revival of German war potential in foreign countries," as a State Department memo inelegantly put it, had been the primary justification for the 1947 decision to make von Braun's temporary stay in America more permanent. Even if the United States didn't need him right now, Washington wanted to make certain no one else got his expertise. Due to their "threat to world security," the Germans couldn't be repatriated. To a frustrated von Braun, it seemed that the United States had dumped him and his team in deepest Texas and forgotten all about them. "We were distrusted aliens living in what for us was a desolate region of a foreign land," he recalled. "Nobody seemed much interested in work that smelled of weapons."

• • •

Von Braun might have languished indefinitely under the hot Texas sun if not for the actions of two men: Senator Joseph McCarthy of Wisconsin and North Korea's Communist leader, Kim Il Sung.

McCarthy's Red-baiting reign of terror—when everyone from J. Robert Oppenheimer to what seemed like half of Hollywood was accused of Communist sympathies—inadvertently helped to rehabilitate von Braun. The only skeletons in his closet were Nazi skeletons, and the Reich was yesterday's enemy. Berlin was no longer the seat of evil. The divided city had been transformed into a symbol of freedom by the massive airlift orchestrated by Symington and LeMay in 1948, when three hundred thousand sorties were flown, delivering food and medicine during Stalin's yearlong siege of West Berlin. The cold war by then had begun in earnest, and a new adversary had replaced fascism as an ideological threat to the American way of life.

The threat was magnified a thousandfold the following year, when Moscow detonated its first atomic bomb. The era, grumbled LeMay, "when we might have completely destroyed Russia and not even skinned our elbows doing it" was over. A few months later, the revolutionary cancer spread to China, further whipping up domestic paranoia in the United States. Suddenly, any American who had ever attended a Marxist meeting in the 1930s, or dated someone who had, was potentially a security risk. Immigrant scientists, with their top-secret clearances and Eastern European backgrounds, were especially vulnerable. After Mao's victory, the blacklist was expanded to include Chinese-born American researchers. Tsien Hsue-shen, one of the founders of the Jet Propulsion Laboratory at Caltech and a pioneer in the field of American rocketry, was arrested and deported to China. (Documents would later reveal that he had been innocent of the spying charges, prompting one military historian to call the affair the "greatest act of stupidity of the McCarthyist period. . . . China now has nuclear missiles capable of hitting the United States in large part because of Tsien.")

Von Braun, however, was above reproach. For five years the FBI and army intelligence had monitored his every move, read his correspondence, and listened to his telephone conversations. Nothing more untoward than the suspect sale of some undeclared silver by his brother Magnus had ever been uncovered. By the summer of 1950, the political climate had changed sufficiently enough for von Braun to come out of purgatory. What's more, his services were finally needed again. On the night of June 25, 1950, Kim Il Sung's North Korean forces crossed the thirty-eighth parallel, threatening to overrun South Korea. This time President Truman decided to draw the line. Three weeks later von Braun received his first meaningful commission, the Redstone.

With the assignment came a change of address and a new lease on life for von Braun and his team. Compared to Fort Bliss, Huntsville seemed idyllic. The historic hamlet was home to fifteen thousand genteel southerners, and its proud Civil War heritage was etched in the

Confederate Monument that crowned the town square. White clapboard church spires dominated the skyline, and the sidewalks were trimmed with white picket fences and immaculately groomed lawns. White was the dominant color in Huntsville, as it was in all of Madison County, Alabama, and throughout the entire Jim Crow South. But the civil rights movement was beginning to take root in 1950, and this worried some of the town's newest German residents. "We had some concerns here," Wernher Dahm recalled. "Not so much about segregation . . . as about open strife."

Huntsville wasn't perfect. But after wandering in the desert for five years, von Braun and his crew had finally found a home. They now had regular salaries, as opposed to the six dollars a day they had received during their first days in Texas, and they had complete freedom of movement. Their legal status had been "normalized," thanks to some creative immigration paperwork, and they would be eligible for U.S. citizenship in a few years. Meanwhile, they could buy cars, houses, motorboats, and televisions; in short, they could enjoy all the material benefits of the American dream.

Life was looking up for von Braun. He bought a fashionable new rambler on a large hilly lot on McClung Street just outside Huntsville's historic city center. His wife, Maria, gave birth to two daughters, Margrit and Iris. The Redstone performed flawlessly during its 1953 tests, and Walt Disney came calling the following year with an intriguing television offer.

Disney was putting together a weekly television program to promote the new theme park he was building on 160 acres of orange groves in Anaheim, California. The show, like the amusement park, would be called *Disneyland*, and each week would feature a theme from one of the park's four attractions: Fantasyland, Adventureland, Frontierland, and Tomorrowland. The fledgling ABC network was backing the venture with $4.5 million in loan guarantees to Disney personally. No one else had wanted to put up the money, not even Disney's brother Roy, who ran the business side of the animation studio. But Walt Disney had run the numbers. Americans had given birth in droves after the war, and their children were now clamoring for entertainment. The country's Gross Domestic Product had almost doubled over the past decade, thanks in part to the industrial ramp-up for war production and America's subsequent competitive advantage over the decimated European economies. The boom had even trickled down to minimum-wage earners, whose hourly pay in 1954 had just been increased from seventy cents to a dollar, giving consumers at the bottom of the scale a little extra to spend on vacations. That year, for the first time in history, airlines had supplanted both rail and ships as the primary mode of transcontinental and transatlantic traffic, making travel less time-consuming. And the explosion of cars on American roads made it more affordable for the masses. For Disney, it all added up to one thing: a tourism boom. All he needed were telegenic hosts for his Sunday night Disneyland infomercials on ABC. The actor Ronald Reagan would handle the grand opening. Was the handsome von Braun interested in hosting the "Man in Space" segments of the Tomorrowland programs?

It was thus from the odd pairing of Mickey Mouse and a former SS major with a Teutonic Texas twang that most Americans first heard of the futuristic concept of an earth-orbiting satellite. Disney himself in-troduced the inaugural Tomorrowland program on March 9, 1955. "In our modern world," he declared, "everywhere we look we see the influ-ence science has on our daily lives. Discoveries that were miracles a few short years ago are accepted as commonplace today. Many of the things that seem impossible now will become realities tomorrow." The Milky Way briefly replaced Disney on the screen. "One of man's oldest dreams has been the desire to travel in space. Until recently this seemed to be an impossibility. But new discoveries have brought us on the threshold of a new frontier."

The show attracted 42 million viewers and made von Braun a star. With his youthful good looks, broad shoulders, and perfect blend of boyish enthusiasm and European erudition, von Braun quickly became America's space prophet, a televangelist leading audiences on the scien-tific conquest of distant planets. The country was enthralled by the endless possibilities unleashed by jet travel and the splitting of the atom. Modernism swept architecture, automobile, and furniture de-sign, where styles were sharp, angular, and edgy. Engine Charlie's Gen-eral Motors incorporated the fast, futuristic themes in its famous Motorama car shows, which featured Delta-wing designs, high fins, and space-bubble taillights. Cool, crisp colors announced the new era: pale greens, baby blues, pastel yellows, and deep, pile-rug whites. Earth tones were out; glass and stainless steel were in. At the movies, aliens reigned as *Invaders from Mars* and *War of the Worlds* dominated the box office. In fashion, the look was tapered, sleek, and hurried. In music, the new sound was rushed, the rapid-fire rhythm of rock and roll. Speed was the essence of the jet age, and it found expression even in the national diet. In 1955, the entrepreneur Ray Kroc franchised his first two McDonald's restaurants. He called them fast-food outlets.

The Disney series had tapped into this headlong dash toward the future and inflamed the imagination of Americans in much the same way that the writings of Hermann Oberth and the films of Fritz Lang had ignited the amateur rocketry craze in Germany in the late 1920s that eventually produced the V-2. But the space craze that swept the nation in the mid-1950s did not translate into political support for space programs. Lawmakers had more earthbound and immediate concerns. Space was the distant domain of dreamers and science fiction writers.

Until 1954, cumulative federal spending on satellite research in the United States amounted to $88,000. And even this sum was thought excessive. When von Braun that same year offered to launch a satellite using a modified Redstone missile for less than $100,000, his request for the additional funding was flatly denied. If money was any indication of political interest, satellite and space exploration did not even register on radar screens in the nation's capital until 1955, when the Soviet Academy of Sciences announced that it would try to send a craft into orbit as part of the planned International Geophysical Year. (The pledge was meaningless, however, until Khrushchev and the Presidium got on board the following year.)

The IGY was a science Olympics of sorts that had its origins in the nineteenth-century Arctic exploration races. Held every fifty years to encourage scientific exchange on the physical properties of the polar regions, it had expanded its brief to cover the planet's skies, oceans, and ice caps. A few weeks prior to the IGY's 1955 convention in Rome, which set July 1, 1957, as the start of the next Geophysical Year, Radio Moscow announced that the Soviet Union would launch scientific instruments into space to measure such phenomena as solar radiation and cosmic rays. In response, the National Academy of Sciences promptly declared that the United States would also send up a satellite to study the earth's protective cocoon. "The atmosphere of the earth acts as a huge shield against many types of radiation and objects that are found in outer space," the academy press release stated, somewhat dryly. "In order to acquire data that are presently unavailable, it is most important that scientists be able to place instruments outside the earth's atmosphere in such a way that they can make continuing records about the various properties about which information is desired."

Behind the seemingly benign facade of the announcements, both the United States and the Soviet Union had ulterior motives for participating in the IGY. Each was looking for a peaceful, civilian excuse to test the military potential of its hardware—how air drag, gravitational fields, and ion content could affect missile trajectories; how the ionosphere affected communications; how orbital decay worked; and so on. There was another, equally important consideration: a research satellite blessed by the international scientific community would set the precedent for an "open skies" policy where sovereign airspace did extend beyond the stratosphere. This, more than anything, argued James Killian at the Office of Defense Mobilization, justified the military's support for the IGY program.

Killian was among only a handful of people in Washington in 1955 who grasped the true significance of a satellite. Engine Charlie Wilson was not one of them. Killian said that the secretary of defense displayed a "narrowly limited understanding" of the new technology and balked at backing the project. Eventually a compromise was struck in which Congress allocated $13 million—roughly the cost of two B-52 bombers—for America's entry into the space race. The funding was hopelessly inadequate, and Nelson A. Rockefeller, another Eisenhower adviser, urged that the administration take satellites more seriously. "I am impressed," he wrote in a widely circulated memorandum, "by the costly consequences of allowing the Russian initiative to outrun ours through an achievement that will symbolize scientific and technological advancement to people everywhere. The stake of prestige that is involved makes this a race we cannot afford to lose." The "unmistakable relationship," Rockefeller added ominously, "to intercontinental ballistic missile technology might have important repercussions on the political determination of free world countries to resist communist threats."

But Eisenhower didn't bite. For him, nuclear bombers were a sufficient stick to keep the Soviets in line. Khrushchev's secret speech, in fact, would be interpreted by the CIA as proof that the existing American doctrine was working. "It must be restated, and it cannot be emphasized too strongly," the CIA said of Khrushchev's break with Stalinism, "that recognition by the Soviet leaders of the significance of nuclear weapons is the underlying cause for their policy shift. For the present, the atom and jet are the basic deterrent."

Bombers had actually scared the Kremlin into softening its stand— "mellowing," as the agency put it. Satellites, therefore, were trivial in the superpower rivalry. Eisenhower declared that he felt no need "to compete with the Soviets in this area." Secretary Wilson told the *New York Times* that he too was unconcerned about the prospect of Russia reaching space first. "I wouldn't care if they did," he said.

5

DESERT FIRES

The R-7's rescheduled March 1957 test-launch deadline came and went, and Sergei Korolev was still not ready. He was becoming increasingly irritable, more prone to terrorizing his staff with his infamous flare-ups, and he wielded both the stick and the carrot to motivate his engineers. Breakthroughs were rewarded with on-the-spot bonuses, wads of rubles that Korolev kept in an office safe for just such a purpose. The most significant achievements earned holidays to Black Sea resorts or even the most sought-after commodity in the entire Soviet Union: the keys to a new apartment. Like all large factories and institutes, OKB-1 was responsible for housing its employees. For young and especially newlywed scientists living in dormitories without privacy, few incentives matched the prospect of skipping to the front of the long waiting list for a place of their own.

The capitalist approach was paying dividends, as glitches were progressively ironed out. Valentin Glushko had wanted no part in making modifications to the steering thrusters, so they were completed in-house at OKB-1. The addition of the tiny directional thrusters also solved the problem of postimpulse boost; the R-7's main engines could now be shut down a fraction of a second early to compensate for residual propellant, while the little thrusters could be fired to adjust speed and position at the critical aiming, or "sweet," point, as guidance specialists called cutoff.

A backup radio-controlled radar guidance system was also installed to augment the accuracy of the onboard inertial gyroscopes. Boris Chertok supervised the duplicate system, which had involved building nine tracking stations deep in the Kazakh desert over the first 500 miles of the R-7's trajectory route, and another six stations within a 90-mile approach to the target area in the Kamchatka Peninsula, 4,000 miles farther. As the missile would fly over the first nine stations, radar would pick it up, plot whether it was on course, and send back telemetry signals for any needed adjustments to trim, yaw, or pitch in what essentially was a very large and costly version of amateur modelers flying radio-controlled airplanes. But with the R-7, the action would unfold at speeds in excess of 17,000 miles per hour and span many time zones in a matter of minutes.

With the R-7 now capable of guided flight, Korolev's team solved the problem of how to support its crushing weight during liftoff. Their solution was ingeniously simple. A gigantic vise with collapsible jaws, pivots, and counterweights was built in Moscow and tested at a military shipyard in Leningrad—the only place with cranes big enough to lift the R-7. When the rocket was lowered into this vise, its 283-ton mass forced the jaws to squeeze shut, like the petals of a tulip. As the rocket rose after ignition, relieving the load from the clamped jaws, the hinged counterweights at the end of the petals exerted downward force, releasing the vise grip.

On the other hand, concerns about the R-7's nose cone persisted throughout the winter and spring of 1957. After much experimentation, the warhead receptacles were blunted and shortened to reduce drag, while an ablative material that used a combination of silica and asbestum with textalyte was selected for the heat shield. The reentry problem, as Korolev would discover, was far from licked, but by late April the Chief Designer felt sufficiently confident that he moved his entire operation to the secret facility that had been built for the R-7 in Kazakhstan to finally take the ICBM out for its long-delayed test flight.

Originally known as Tyura-Tam and eventually as Baikonur (in yet another instance of Soviet misdirection), the installation was one of the most closely guarded secrets in the Soviet Union. Correspondence to it was simply addressed to Moscow Post Office Box 300, and it figured on no map. The military construction crews that had built it had been rotated in short, highly compartmentalized shifts, and never told what they were doing in the middle of the broiling Kazakh desert, where temperatures soared to 135 degrees in summer and plunged to 35 below in the short, merciless winter. Dust storms clogged machinery with salty sand particles blown across the central Asian steppe from the Aral Sea. For hundreds of miles, there was not a single tree to offer shade or firewood. Soldiers slept in tents crawling with scorpions, and water had to be trucked in.

Besides the benefit of seclusion, the Tyura-Tam site had been selected because of the R-7's guidance problems. Tracking stations could be built in the depopulated area along the route to Siberia, and rocket engineers didn't have to worry about spent boosters falling on urban centers. A major rail spur also happened to run nearby, easing the transport of missile parts and building materials. The pace of construction at Tyura-Tam had been so frenetic, so filled with accidents and setbacks, that the young military officer responsible for erecting the Soviet Union's first nuclear ICBM launch site had gone crazy and been carted away to a psychiatric ward. Only the rudimentary infrastructure had been completed in time for testing: a huge assembly hangar connected by rail to a fire pit the size of a large quarry to absorb the flames at liftoff; several underground propellant storage tanks; a cement bunker blockhouse; and a 15,000-square-foot launch platform erected out of sixteen massive bridge trusses. Otherwise, and in every direction, there were only shifting dunes and the shimmering heat waves of the sun reflecting off the sand. There had been no time to build housing, cafeterias, recreation halls, laboratories, or even lavatories. Scientists lived four to a compartment in train cars that had been built in East Germany after the war and equipped with special labs for the storage and testing of delicate rocket components. Showers were a rare luxury. The food was gritty and abysmal. And at night the principal form of entertainment seemed to involve watching scorpions duel to the death in empty vodka bottles.

Ill omens appeared almost from the start. During a dress rehearsal for the first launch, a fire alarm was inadvertently triggered, setting off sprinklers in the blockhouse control room that drenched everyone, including irate military representatives who had come to observe the birth of the Red Army's latest and most lethal weapon. Later, a technician dropped a bolt inside the R-7, and for several tense hours the missile was searched for the lost object. Korolev exploded when Chertok informed him that one of the Moscow dignitaries had offered the clumsy technician a bonus of 250 rubles—about a week's pay—for owning up to the potentially catastrophic mishap. "What the hell are you doing?" the Chief Designer roared. "You should punish him, not reward him. Rescind the bonus immediately and issue a reprimand."

As the May 15 countdown neared, tension grew. A few hours before liftoff, Korolev blew up at another military observer, Colonel Alexander Maksimov, who he thought was sending false reports about leaky liquid oxygen feed lines to his superiors. "Get him out of here right now, or I'm aborting," Korolev shouted. Only it turned out that Maksimov was innocent, and Korolev had to apologize publicly. Any leaky valves had been fixed; Korolev's head of testing, the equally crotchety Leonid Voskresenskiy—the only person among the thousands of NII-88 and OKB-1 employees permitted to address Sergei Korolev by his first name, without the formal patronymic—had a decidedly low-tech method for dealing with leaks. He would wrap his cap over the faulty valve and urinate on it. The minus-297-degree liquid oxygen would freeze the urine on contact, sealing the leak.

The rocket was cleared for liftoff just after 7:00 PM Moscow time on May 15. The massive engines fired, the Tulip launch stand worked perfectly, and the R-7 rose to the whooping cheers of assembled engineers and VIPs. But ninety-eight seconds into the flight something went terribly wrong. The missile crashed, scattering debris over a 250-mile radius.

Back in Moscow, Sergei Khrushchev recalled the phone ringing after dinner at the family's mansion in Lenin Hills. It was the white line, the German-made phone used for important government business. "That was Korolev," said Nikita Khrushchev, looking particularly gloomy as he hung up. "They launched the R-7 this evening. Unfortunately it was unsuccessful."

Korolev, though, was upbeat. First launches almost always failed. That had been the rule with the R-1, the R-2, and the R-5. And he had reason to feel optimistic: for the first minute and a half, the R-7 had performed flawlessly. The trouble had been with one of the peripheral boosters, block D, which had caught fire shortly before separation. The suspected cause of the explosion was excessive vibrations, what was known as the Pogo effect. Korolev and his team would figure out exactly what had happened and fix the glitch. Next time, he was certain, his missile would make it all the way to the target zone in Kamchatka, almost 5,000 miles away on Siberia's Pacific coast.

For a long, hot month, they tinkered. At last, on June 9, another R-7 was wedged into the Tulip launch stand. Voskresenskiy supervised all the preparations. Korolev trusted him implicitly, and on more than one occasion he had proven his loyalty and courage. Once, when a launch had misfired and the live warhead had been dislodged from its missile, dangling precariously over the pad, everyone had frozen in panic. But Voskresenskiy had calmly told Korolev, "Give me a crane, some cash, five men of my choosing, and three hours." With wads of vodka-walking-around money bulging out of their pockets, Voskresenskiy's men safely dismantled the one-ton warhead, after which they got royally drunk.

Like a great many test pilots and other people who push safety envelopes for a living, Voskresenskiy was deeply superstitious. So when the next R-7 failed to start, not once, not twice, but on three consecutive days before sputtering out with a smoky cough on the launchpad on June 11, Voskresenskiy decided it was cursed. "Take it away," he ordered. "I never want to see it again." The blighted rocket was hauled away in disgrace.

A dark, defeated mood settled over the exhausted R-7 team. They hadn't seen their families in several months. They were working round the clock, seven days a week, and people were getting sick from the long hours and unremitting worry. Chertok came down with a strange ailment with similar symptoms to radiation poisoning. Eventually he would have to be medically evacuated to Moscow. Korolev developed strep throat and had to take frequent penicillin shots. His health had never fully recovered from the ravages of the camps, and he frequently took ill. "We are working under a great strain, both physical and emotional," he wrote his second wife, Nina. "Everyone feels a bit sick. I want to hug you and forget about all this stress."

Korolev had met Nina Kotenkova at OKB-1, where she served as the institute's English-language specialist, translating Western scientific periodicals. It was through her that the Chief Designer had kept abreast of Wernher von Braun's writings and exploits in the American media, and in the flush of lonely nights spent jointly hunched over the pages of *Popular Mechanics* they had fallen in love. Korolev had still been married to his first wife, Ksenia, when they had met, and their affair had led to a bitter divorce and estrangement from his teenage daughter, Natalia, who for years refused to see him and would never agree to speak to Nina.

From his monastic hut at Tyura-Tam, a cabin without running water and only a bare lightbulb to illuminate the lonely gloom, Korolev wrote his daughter weekly, begging for her forgiveness. He tried calling on her twenty-first birthday, but she hung up on him. "It hurts me so much," he told Nina.

Even at the best of times, Tyura-Tam was a dispiriting place. But with two consecutive failures, growing friction between the different design bureaus over responsibility for the myriad malfunctions, and Moscow becoming increasingly irritated, the atmosphere was downright foreboding as Korolev lined up a third R-7 for launch on July 12. This time the countdown went uninterrupted, the engines all fired properly, and the missile lifted off without a hitch at 3:53 PM. Thirty-three seconds later it disintegrated. The strap-on peripheral boosters had separated early.

Watching the four flaming boosters slowly sail down into the desert just four humiliating miles from the launchpad, Korolev dejectedly shook his head. "We are criminals," he said. "We just burned away [the financial equivalent of] an entire town."

Fear now descended on the despondent scientists. The Soviet Union was not so far removed from the Stalinist era to presume that punishment might not be meted out for the catastrophic and costly failures. "What can they do to us?" a frightened Chertok asked Konstantin Rudnev, the deputy minister of armaments. "They are not going to jail us, or send us to Kolyma?"

"No," Rudnev reassured him. "But your rocket will be put into the hands of other people." He reminded Chertok that Korolev was not the only designer who had caught the Kremlin's eye. There was the talented Vladimir Chelomey over at OKB-52, where the young Sergei

Khrushchev was about to go to work. "He was a brilliant scientist, but alienated people," Sergei recalled. At OKB-586, Mikhail Yangel had just landed a commission to build the 1,200-mile-range R-12 missiles that Khrushchev would later try to ship to Cuba. Yangel, especially, was viewed by many as the complete package: a gifted engineer, an astute manager, and an effective salesman. Korolev's main strengths were his ability to sell Khrushchev on his ideas and his hard-driving, almost maniacal management style; he was not seen as technically brilliant. In fact, there were dark rumblings that the dozen additional mini–steering thrusters that he had grafted onto Glushko's engines were actually causing some of the malfunctions. They were "worthless," Glushko railed, and a clear demonstration that the arrogant Korolev had overstepped his "technical competence."

It wasn't just Glushko's increasingly vocal recriminations that the Chief Designer had to worry about. He also had powerful enemies in Moscow, including the ruthless Central Committee official Ivan "the Terrible" Serbin, who was jealous of his independence and status as Khrushchev's pet designer. Many in the military were equally resentful and begrudged the vast sums Korolev's missiles diverted from conventional forces. (One general had famously groused that if the alcohol wasted on rocket propellant were given to his soldiers instead, they could wipe out any target more effectively.) The Soviet brass was by no means enamored of rockets. "You can't count on Malinovsky," Rudnev warned Chertok, referring to the deputy defense minister, Rodion Malinovsky. "He only tolerates you because Khrushchev supports the rocket."

But how long would the support last? Already there were mutterings to stop wasting money on R-7 flight tests. "The rocket will fly," Korolev stubbornly maintained. But others were starting to have their doubts. In Moscow a whispering campaign had begun that Khrushchev had backed the wrong horse, that once more his rash decisions were endangering the empire. Without the R-7 as a deterrent, the Soviet Union had horribly exposed itself by cutting conventional forces. What would stop the madman LeMay, and his huge armada of new B-52s, from laying waste to Russia now?

If there was one consolation, one sliver of good news, it was that the American missile effort seemed to be faring even worse. There were encouraging reports of explosions and misfires plaguing the Thor and Atlas trials. Von Braun's Jupiter program was on the verge of being canceled, and its first test launch had ended in a spectacular fireball. Most heartening of all was word that the American secretary of defense had just announced his intention to cut $200 million from rocket spending. Wilson was putting a freeze on all new missile proposals and had banned overtime on existing projects.

Though it didn't appear that the U.S. missile program was going to overtake the Soviets anytime soon, for some reason Korolev was fixated on the notion that von Braun was going to launch a satellite at any moment. The Americans are on the verge, he kept telling anyone who would listen, and he obsessively scoured Nina's translations of Western publications for any hint that might betray America's orbital intentions.

The Chief Designer's manic paranoia on the subject had begun to irk his exhausted colleagues at Tyura-Tam, where tempers were flaring and the blame game among the designers was reaching new heights. They had now had five failures, and everyone was pointing fingers.

"And you and your rocket? Are they not to blame?" Glushko shouted at Korolev during a particularly heated exchange. "What about the draining [valves]? What if the feed line erupted?"

"You should understand," Korolev shot back. "There are no Korolev rockets. These are our rockets with your engines, with his radio guidance," pointing to Nikolai Pilyugin from the NII-885 design bureau. "Your approach to this case has been flawed from the beginning," he went on. "Yes, the rocket can fail because of his launch pad"—Korolev nodded toward Vladimir Barmin, the man behind the Tulip—"because of the failure of your engines, because of the failure of his equipment"—Korolev's glare now fell on Viktor Kuznetsov, the inertial gyroscope expert—"or because of my drainage valves. But each time is a failure of our rocket. We all should be responsible."

Vasiliy Ryabikov, the chairman of the State Commission on the R-7, the Kremlin's direct representative, wasn't buying it. "You are a very cunning person, Sergei Pavlovich," he said. "You spread so much stink on others while perfuming your own shit."

Marshal Mitrofan Nedelin also seemed to have lost confidence in the once-golden Korolev. The head of the Soviet Union's strategic rocket forces now wanted testing stopped.

"I agree with Mitrofan Ivanovich," Glushko piled on, knowing that his opinion carried far more weight than all the other subcontractors. His bureau had a virtual monopoly on Soviet rocket engine manufacture, and he supplied all of Korolev's competitors. Without his motors, there was no Soviet missile program or space program. "There's no reason to continue these tests," he huffed. "[Fifteen] of my excellent engines are destroyed, and if it goes on this way, my production line won't be able to keep up."

"But Valentin Petrovich," Konstantin Rudnev addressed Glushko, in Korolev's defense, "if the rockets flew according to schedule your engines would be destroyed anyway."

"I wouldn't begrudge the engines if they served their purpose," Glushko retorted. "But why should I suffer from somebody else's mistakes?"

"This is not *somebody's* fault," Korolev exploded. "It is *our* fault."

Despite Korolev's attempt to spread the bureaucratic blame, the message from Moscow was clear: it was his rocket and his responsibility, and the State Commission on the R-7 was getting ready to recommend pulling the plug. Korolev's career, his dreams, his future—possibly even his freedom—hung by a thread. He was alone, literally sick and tired, stuck in the hellish Kazakh desert, and the vultures were circling. For the supremely self-confident Korolev, it was an unaccustomed and frightening predicament. "When things are going badly, I have fewer 'friends,'" he wrote Nina with uncharacteristic humility. "My frame of mind is bad. I will not hide it, it is very difficult to get through our failures. . . . There is a state of alarm and worry. It is a hot 55 degrees [131 Fahrenheit] here."

• • •

Sergei Pavlovich Korolev did not grow up dreaming of rockets or the stars. Space had never captivated the stocky, solitary youngster during the formative years he spent in lonely isolation at his grandparents' estate in prerevolutionary Ukraine. "Sergei was about three when our family disintegrated," his mother, Maria Balanina, recalled. A bitter divorce and prolonged custody battle had split the Korolevs in 1911, and Maria, an unusually headstrong woman, had left her only child with her parents while she completed her university degree. Few women of the era pursued higher education, and Maria, a dark-haired beauty with porcelain features, a wasp waist, and impressively plumed bonnets, juggled her dueling parental and academic responsibilities with a fiery determination she would pass on to her famously obstinate son.

The two lived mostly apart. During the week Maria studied French and literature at the Ladies College in the Ukrainian capital of Kiev, while nannies and tutors cared for Sergei. On weekends she made the two-hour trip home to Nezhin, a small town along the bustling trade route that linked the czarist and Hapsburg empires. Her parents were wealthy merchants; in a photograph, one of their two stores sits at the foot of a four-spire church, occupying a low, block-long structure that resembles a turn-of-the-century strip mall. As a side business, the family had a small but highly successful brine operation that had applied to receive the coveted imperial seal as the official purveyor of pickles to the court of Czar Nicholas II.

Though young Sergei did not want materially, he lacked companionship and the freedom every child desires to roam. "He didn't have any friends of his own age, and never knew children's games," his tutor, Lidia Mavrikievna Grinfeld, recalled. "He was often completely alone at home and . . . would sit a long while on the upper cellar door and watch what was happening in the street." Moreover, the gates to the family estate were always locked because Korolev's grandparents worried that his estranged father, whom he was not permitted to see, would try to kidnap him. "I felt I needed to keep him at home," Maria later explained. "He was so impressionable and thin-skinned and I had to teach him how to better cope with reality."

Locked away in his splendid isolation, Korolev built giant dollhouses and cried frequently. But the seclusion apparently instilled in him a self-reliance and vigorous imagination that would serve him well later in life. In the summer of 1913, when Korolev was six, an event occurred that would leave a lasting imprint on the melancholy child. "A poster appeared on the market square that announced that Pilot Utochkin would perform a flight for the public," Maria recalled. No one in Nezhin had ever seen an airplane before; automobiles, at the time, were rare. "People were very excited. Some didn't believe man could fly like a bird. The entire city turned out for the spectacle."

Perched on his granddad's stout shoulders, little Sergei watched rapt as the small four-winged contraption careened down the dusty fields, bouncing fifty feet in the air before wobbling back to earth near a convent a few miles away. The sight of such a display of freedom stirred a powerful urge in the sheltered young boy, and he couldn't stop "babbling" when he got home.

"Mother, can you give me two new bed-sheets?"

"What for?"

"I will tie them to my arms and legs and climb to the top of the smokestack and fly."

"You'll kill yourself."

"No, birds can fly."

"But birds have rigid wings."

Thus, according to family legend, was Korolev's passion for flight ignited. His juvenile aspirations, though, were still relatively modest and did not yet soar beyond the clouds: he simply wanted to be a barnstorming pilot, a dashing daredevil with a white, flowing scarf. Perhaps it was the swagger, the star status, and the supreme self-confidence of early pilots like Utochkin that left such an indelible mark on a fatherless child desperately yearning for a male role model, the character traits that would form the bedrock of Korolev's adult personality.

In 1916 Sergei acquired both a father figure and the opportunity to nurture his growing obsession with aircraft. Maria remarried that year, to a kind and gentle railway engineer, and the family later relocated to Odessa, where Korolev's new stepfather, Grigory Balanin, had been appointed to a senior position at the harbormaster's office. Odessa was a rough-and-tumble port city of palm trees and prostitutes, smugglers and sailors. It had always enjoyed an exotically lawless reputation in czarist times as an entrepreneurial haven with an unusually cosmopolitan makeup; Jews, Armenians, Greeks, Italians, Russians, and Ukrainians mingled easily with the resident representatives of virtually every Mediterranean seafaring culture. The city changed hands several times during the Bolshevik revolution, and a French detachment supporting the czar's White Army was still stationed there when Maria, Sergei, and Balanin moved into a three-bedroom apartment overlooking the Black Sea. Reds, Whites, and Ukrainian nationalists began block-by-block battles for control of the ravaged town. Korolev's school closed during the worst fighting, and food was in short supply. "Hunger, chaos, a city filled with refugees," Maria recalled. "There were homeless children living in lobbies and courtyards, and the authorities changed frequently."

But there were also military aircraft, a squadron of plywood hydroplanes enticingly anchored within view of Sergei's balcony, separated from the harbor traffic by a fence of barbed wire. For the young Korolev, the proximity was irresistible. He would swim out past the jetty, his mother wrote, and "hang onto the barbed wire for hours, as if mesmerized, watching with interest what was going on there. Once a mechanic shouted to him: 'Well, what are you hanging around for? Why don't you give me a hand? Can't you see I'm having difficulty with this motor?' That was all Sergei needed. He quickly crawled under the wire. Soon they got used to seeing him around in the detachment."

When the civil war finally ended in 1921 and classes resumed, Korolev found himself drawn to mathematics and drafting, a subject introduced by the new Soviet government. Still shy from the years of home schooling and the disruptions of war, Sergei did not socialize much. "He was not interested in small talk like the other kids," according to his daughter, Natalia. Sports were never really interesting to him either; he joined the gymnastics team only because "he felt it was important to stay fit to become an aviator," said his daughter.

Discipline seemed to come easily to the fifteen-year-old future Chief Designer. "6:00AM Rise," he recorded in his 1922 daily planner. "6:15: calisthenics; 6:30: Breakfast; 7–8:00AM Swim in the Sea; 8:30–1PM: School . . ." The morning swim, of course, was a euphemism for hanging out at the seaplane base, where by now he had become such a fixture that the pilots took him up regularly for rides. Prudently, he did not share this bit of potentially unsettling intelligence with his mother, though Maria suspected her son was up to something after he blurted out one day, "Oh Mother, if you could only see the clouds from the top."

It was in his senior year in 1924 that Sergei Korolev finally began to bloom. He joined an amateur aviation club, started to come out of his shell, and developed a belatedly healthy interest in girls. He was particularly smitten by a classmate of Italian ancestry, Ksenia Vincentini, a fiery brunette who would become a prominent surgeon and his first wife. Korolev also designed his first airplane as part of his drafting class graduation project, a glider that he ambitiously called the K-5, as in Korolev Five. The four earlier versions were presumably little more than doodles, but the design was good enough to be chosen by the Ukrainian Society of Aviation and Aerial Navigation for construction. "That was the definitive moment for Sergei," Maria proudly remembered, "when he chose his career."

Aeronautical engineering was still a relatively new field in 1925, and Korolev enrolled at the Polytechnical Institute of Kiev, which produced such graduates as Igor Sikorsky, the future helicopter designer. In Kiev, Korolev entered a small and obsessive community of aircraft builders, reveling in the heady, hands-on atmosphere, working late into the night and on weekends. He designed and built another glider, which he flew himself, and by his sophomore year his grades were good enough to transfer to the more prestigious Higher Technical School in Moscow, where his mother and stepfather had just moved from Odessa.

In Moscow, Korolev studied under the great Andrei Tupolev, who was already emerging as Russia's most prolific designer of large-frame aircraft. Under Tupolev, Korolev designed his first motorized cub plane as a graduation project in 1930. It was an ungainly snub-nosed craft with a squat twenty-two-foot fuselage and a top speed of one hundred miles per hour. He called it the SK-4 and proudly painted a dark racing stripe down the side of the prototype. Alas, the SK-4 crashed on its third flight. "To my dear friend Piotr Frolov," Korolev inscribed a photo of the wreckage to a fellow student, "in memory of our joint collaboration on this unhappy machine."

The SK-4's technical shortcomings showed Korolev's limitations as a designer. He did not have the artistry, flair, or intuitive vision of others in his graduating class, and his first real job, working on hydroplanes, was not a particularly plum assignment. What Korolev did possess, however, was an uncanny knack for spotting talent, which he did during a chance encounter at a glider club outing in October 1931, when he met two rocket enthusiasts, Friedrich Tsander and Mikhail Tikhonravov. Tsander was a Latvian of German extraction, twenty years older than Korolev, and the founder of a rapidly growing volunteer rocket association called the Group for Studying Reaction Propulsion, or GIRD, whose branches would spread to ninety Soviet cities. Tsander had been a disciple of Konstantin Tsiolkowsky and the author of a popular tract

on interplanetary travel. Eloquent and obsessed, sickly and impover-
ished, Tsander had a Rasputin-like hold on a legion of young Soviet scientists, who pooled funds to support his research.

Tikhonravov, on the other hand, was shy and unassuming—his name
fittingly translates as Quietman—and he was not blessed with the
ephemeral qualities that make inspirational leaders. But he possessed
acute faculties, a reputation for deep thoughts, and "the air of a man
who had already sampled the mysteries of another planet," in the words
of the British historian Deborah Cadbury. Classically trained in French
and Latin, Tikhonravov would coin the term *cosmonaut*, Latin for "space
traveler," leaving the United States to settle for the slightly less accurate
astronaut, or "star traveler," to distinguish its spacemen, though stars, of
course, could not be traveled to.

Tsander in 1931 had been working on a small rocket engine, and
Korolev hit upon the idea of grafting it to a tailless, trapezoidal glider
that he had used from time to time while training to qualify for his pilot's license. The suggestion marked the first hint of where the future
Chief Designer's real talents lay: as an organizer, pulling together other
people's inventions. It was also his first spark of a dawning realization
that rockets could have immediate and practical applications. Attached
to wings, they could assist heavily loaded bombers to take off from
short runways.

Korolev threw himself into the project with his customary vigor,
taking only a day off to marry Ksenia, his high school sweetheart from
Odessa, in a rushed civil ceremony that set the low-priority tone for
their unhappy union of competing careers. After a few hurried toasts,
his bride boarded a train back to medical school in Ukraine, and Korolev returned to working on planes by day and rocket motors by night.

Tsander died of typhus in 1933, but by then Korolev was hooked,
spending all his spare time with Tikhonravov, who would become his
lifelong collaborator. That same year, the pair launched Russia's first
liquid propellant rocket, the GIRD-09. It weighed forty-two pounds,
flew 400 yards, and attracted the attention of the military. By the time
of Natalia's birth in 1935, Korolev's hobby had become a profession.
The Soviet authorities had created the Reaction Propulsion Institute,
or RNII, to study the development of missiles, and Korolev was appointed senior engineer.

Then, tragedy. The Great Terror of 1937–38 brought mass arrests and murders, denunciations and deportations. Tupolev and Glushko were imprisoned and charged with sabotage. Korolev's immediate bosses at RNII were executed; he himself was tortured and handed a ten-year sentence. Told that her father was a fearless pilot away on an important mission, Natalia Koroleva to this day vividly remembers her first memory of meeting her dad. It was at the feared Butyrka prison in 1940, under the supervision of an NKVD secret police guard. "But Father, how could your plane land in such a small courtyard?" she asked.

"Little girl," the guard interrupted with a laugh. "It's very easy to land here. Taking off again is much harder."

• • •

Korolev, in the summer of 1957, was not the only one facing serious problems, whose career and possibly freedom were on the line. For Nikita Khrushchev, the chickens were also coming home to roost. The trouble, this time, began innocently enough with a telephone call. Khrushchev was having lunch at his official residence in Lenin Hills on June 18 when the special government hotline rang. Nikolai Bulganin was on the other end. "Nikita, come to the Kremlin," he said, according to Sergei Khrushchev. "We're having a session of the Presidium."

Sergei Khrushchev recalled being struck by the unusual timing. "The weekly meetings were always held on Thursdays and this was a Tuesday," he explained. His father also thought the sudden scheduling change strange. "Nikolai, what's the hurry?" he asked, puzzled. Bulganin muttered something about going over a speech for the upcoming 250th anniversary of the founding of St. Petersburg, or Leningrad, as the city had been renamed. "We can do that on Thursday," said Khrushchev dismissively. But Bulganin persisted.

The last speech Khrushchev had made in Leningrad in May had caused quite a stir in the Presidium. Speaking off the cuff, and without prior consultation of his fellow Presidium members, Khrushchev had predicted that the Soviet Union would overtake the United States in meat and dairy production by 1960. "We will bury you," he had roared in the heat of passion, choosing his words callously, because the world press interpreted the boast not as an agricultural duel but as a threat of nuclear annihilation. Even without the unfortunate reference to mass

graves, the challenge was a tall order given that American farmers produced almost three times as much meat per capita as their Soviet coun-
terparts. Apparently, Khrushchev's competitive spirit, the same insecure
desperation to upstage the Americans that Korolev had played upon,
had gotten the better of him. But Vyacheslav Molotov and Lazar
Kaganovich had fumed that Khrushchev had thrown down a rash challenge that put the USSR in a potentially embarrassing bind. There was
simply no way the creaky collective farms could triple their current
quotas, and the exuberant Khrushchev had set up the Soviet Union
to fail.

"Who's there?" Khrushchev finally asked, growing uneasy. He had
been around palace intriguers long enough to sense that something
was up.

"Everyone who's having lunch," said Bulganin evasively.

The attack on Nikita Khrushchev began the moment he walked into
the gilded conference room in the Kremlin's main administrative building. A dozen pairs of hostile eyes followed his progress as he pushed past
the tall padded door and made his way along the intricately laid parquet
floor to his customary seat at the head of the long green baize table. Before he could sit down, Georgi Malenkov, a former prime minister whose
demotion to minister of machine building Khrushchev had orchestrated
during the post-Stalin power struggle, rose to speak. Khrushchev, he declared, should not chair the meeting. The extraordinary Presidium session had been convened to address his outrageous behavior, and it would
be wrong for Khrushchev to preside over the discussions. Shaking with
rage and slamming the table with such vigor that drinks reportedly rattled, Malenkov launched into a tirade outlining "error after error" and
then nominated Bulganin to take Khrushchev's place.

Khrushchev was stunned. For all his finely tuned political instincts,
his decades of climbing the party ranks, and the considerable survival
skills he had honed at Stalin's court, the coup had taken him completely
by surprise. And this was a coup, there was little doubt. It followed the
exact same script Khrushchev had himself written four years earlier to
get rid of Beria: the sudden session, angry accusations, arrest. Any moment now, Khrushchev could expect some ambitious KGB or army general to burst into the room with handcuffs. That was how Georgy

Zhukov had earned his promotion first to deputy defense minister and then to defense minister and had received candidate Presidium membership. Now Zhukov's predecessor, the crusty Stalinist soldier Kliment Voroshilov, declared Khrushchev "unbearable" and unfit to be party leader. Kaganovich called him a cow "knocking about the whole country" with his rash economic and de-Stalinization policies. When the young Leonid Brezhnev—a candidate Presidium member Khrushchev had rescued from the relative obscurity of the Naval Political Department—rose in his patron's defense, Kaganovich turned viciously on him. "Leonid Ilyich barely had time to utter the first words," Sergei Khrushchev recalled. "Kaganovich, his mustache bristling, loomed over him. The last words Brezhnev heard were something like this: 'You like to talk. You've forgotten how you were relegated to the [NPD]. We'll chase you back there soon enough.' Leonid Ilyich faltered, started to clutch the back of his chair, and sank slowly to the floor. A doctor was summoned. Guards carried the unconscious [Brezhnev] to an adjacent room."

The insults continued, even from within Khrushchev's own camp. "You've become the expert on everything—from agriculture to science to culture," charged Dimitri Shepilov, the *Pravda* editor Khrushchev had appointed foreign minister to replace Molotov, who had held the post for nearly three decades. "Someone who's illiterate can't govern a country."

Khrushchev was ousted as first secretary of the Communist Party by a margin of seven to three, excluding his own vote. But the expected arrest never materialized. For three days and three nights he stayed in the Kremlin waiting for the final ax, as the hard-liners celebrated. But the coup leaders had made a critical mistake: they had not arrested Khrushchev. "They couldn't," explained Sergei Khrushchev, "as long as Father retained the loyalty of two key people: Zhukov and KGB chief Ivan Serov. He had appointed both of them, and they both knew that if he was replaced so would they." Without the backing of the military and the secret police, Molotov and the other conspirators could not risk jailing Khrushchev. "I'm sure they planned on doing so later," Sergei Khrushchev said, "once they were in complete control."

The delay bought Khrushchev badly needed time. He knew that

with the exception of the hard-liners, most members of the Central Committee backed him. He was a man of the people who traveled widely in the provinces, unlike his Stalinist cronies, and most rank-and-file representatives of the Communist Party approved of his liberal reforms. If he could just reach them. With Serov's help, Khrushchev hatched a countercoup, and the two began secretly mobilizing the three hundred elected members of the Central Committee. In theory, only a full plenum of the Central Committee could override the Presidium. In practice, it had never happened before, and it posed logistical problems. Central Committee members were scattered across the Soviet Union, often in inaccessible provinces in Siberia and central Asia, and getting them all to Moscow on Aeroflot passenger planes could take days, and in some cases weeks. Only the KGB could contact them surreptitiously, without alerting Kaganovich's forces. And only Zhukov's new long-range jet bombers could fly them back to the capital in time to make a difference.

By the late afternoon of June 20, forty-eight hours after the coup began, Zhukov had managed to deliver eighty-seven Central Committee members to Moscow. His bombers were landing, refueling, and taking off to pick up more Khrushchev supporters. The new arrivals demanded that the shocked Presidium plotters convene a full party plenum to discuss the leadership crisis. Lest the conspirators forget where the military stood on the matter of Bulganin versus Khrushchev, the delegates descending on the Kremlin were led by a parade of generals and marshals. By June 22, a weeklong plenum had been convened, and Kaganovich, Bulganin, and Molotov were in full retreat. It was now the war hero Zhukov, the savior of Moscow and conqueror of Berlin, who led the countercharge. The coup plotters, he said, were the very men who had been Stalin's bloodiest henchmen, responsible for the worst of the purges. During a murderous eight-month rampage in 1938 alone, he alleged, Kaganovich, Molotov, and Malenkov had personally signed 38,679 execution orders. So had others, they protested. So had Khrushchev. But their tone was defeated. The fight in them had gone. Kaganovich, as Khrushchev noted in his memoir, no longer "roared like an African lion."

Seated once more in the first secretary's chair, Khrushchev could not suppress a satisfied sneer as his opponents squirmed for mercy. The attempted coup had been foiled. All that remained was for Khrushchev to decide the fate of the conspirators.

• • •

Sergei Korolev hadn't had reason to laugh for a long time. But he was in unusually high spirits on the August morning he visited Boris Chertok at Moscow's Burdenko military hospital. "Okay, Boris," he cheerfully chided Chertok. "You continue playing sick, but don't stay out for too long." A half dozen jubilant engineers crowded around Chertok's bed, teasing and poking their infirm colleague. The suspected radiation poisoning had turned out to be simply an exotic Kazakh bug that manifested similar symptoms. "This is the best medication," assured Leonid Voskresenskiy, the daredevil chief of testing, pulling a bottle of cognac out of a bag. "So," he said, once Korolev had finished his pep talk and excused himself on the grounds of an important meeting. "Here's the pickle we're in. Everyone congratulates us, but nobody other than us knows what's really going on."

The rocket, as Chertok already knew, had finally worked. Korolev had been given one final chance to prove himself, and at 3:15 PM on August 21, the R-7 had flown all the way to Kamchatka, landing dead on target next to the Pacific Ocean. Korolev had been so relieved, so euphoric, that he had stayed up till 3:00 AM the next morning, jabbering away excitedly about the barrier they had just broken. There was a slight hitch, however. The heat shield had failed, and the dummy warhead had been incinerated on reentry. Apparently, the nose cone dilemma hadn't been solved after all. And without thermal protection, the R-7 was not an ICBM, just a very large and expensive rocket. That was why Korolev had just rushed off to meet with a group of aerodynamic specialists: to see if they had any solutions to what Voskresenskiy called "Problem Number One."

Telemetry readings had shown that not only had the dummy warhead completely burned up ten miles over its intended target; it had also been rammed from behind in outer space by the main stage on separation. The bumping could be easily solved by venting some of the compressed nitrogen from the fuel tanks to slow the central block at the time of the warhead's ejection, but the thermal shield failures had everyone stumped.

"We've only got one rocket left, number 9," Voskresenskiy continued. "And we don't know what to do to fix the nose cone." The next test launch was scheduled for September 7, and it would have identical results. The R-7 would perform flawlessly; the dummy warhead would be completely destroyed on reentry, confirming the missile's current uselessness as an ICBM and weapon.

"We need to take a break to make radical improvements on the nose cone," Voskresenskiy told Chertok, leaning closer and lowering his voice to a mock-conspiratorial whisper. "While we're working on it, we'll launch satellites. That'll distract Khrushchev's attention from the ICBM."

6

PICTURES IN BLACK AND WHITE

On the morning of August 28, 1957, the same day that Boris Chertok and Leonid Voskresenskiy would sip cognac and swap conspiratorial jokes at the Burdenko military hospital in Moscow, ground crews at a secret airstrip outside Lahore, Pakistan, readied a mysterious plane for takeoff.

The black, single-engine craft bore little resemblance to anything that had ever taken to the skies before. As with a glider that had been retrofitted for powered flight, its slender silhouette defied conventional design. The wings were disproportionately elongated and dog-eared at the tips. Strange metallic poles held them upright like overgrown pogo sticks. An alarmingly slim sail rose from the tall tail section, which seemed so frail that it might crumble at the slightest crosswind. The landing gear was equally unusual and flimsy and appeared to consist of a lone bicycle wheel.

In the predawn Punjabi gloom, the misshapen plane looked all the more alien against the sweltering backdrop of ancient battlements and mosques and the muezzin's call to prayer that echoed softly from minarets in the surrounding hills. Though the heat index had already crossed an oppressively humid one hundred degrees, and shimmering waves would soon rise with the sun from the steaming tarmac, technicians in sweat-stained coveralls pumped an antifreeze additive into the plane's huge fuel tanks. The specially blended gasoline was necessary because it was cold where the aircraft was headed, the coldest place imaginable. But the mechanics gingerly filling the tanks from portable fifty-five-gallon oil drums did not know where that was; their security clearances went no further than maintenance.

While the ground crews made their preflight rounds, another group of technicians, distinguished by their white gloves, fidgeted in a bay under the single-seat cockpit. There, next to three small-diameter portholes that contained the most sophisticated photographic lenses ever devised, they loaded a 12,000-foot-long spool of high-resolution Kodak film. The custom-made film and 500-pound Hycon camera were the only outward clues as to the plane's true purpose. Otherwise, it had no markings, identification numbers, or insignias. No running lights winked under its dark fuselage, which had been painted dull black to better blend in with the night sky. Nowhere in its equally anonymous innards was there a manufacturer's seal or anything else that would betray that the CL-282 Aquatone had been assembled at the top-secret Lockheed Skunkworks plant in Burbank, California.

Officially, the CL-282—or the U-2, as it would eventually be called—did not exist. Neither did the pilot, E. K. Jones, who was going through his own preflight routine in a small, barrack-style building near the runway. Like the twenty-man Quickmove mobile maintenance team and the fuel drums they had brought with them, Jones had been flown in the day before from the U-2 main staging base in Adana, Turkey, to minimize American exposure to prying Pakistani eyes. At 4:00 AM a doctor measured his temperature, pulse, and blood pressure and examined his ears, nose, and throat for signs of infection. The medical exam was a formality; like every one of the two dozen U-2 pilots on the CIA's payroll, Jones was in excellent physical condition. Photographs of the era show a compact, muscular young man, with thick, dark hair and the slightly swaggering expression common to fighter aces, test pilots, and other alpha males of the airborne community. Pronounced fit for duty, Jones made his way to a tiny cantina, where cooks had prepared a high-protein meal of steak and eggs to sustain his stomach during his nine-hour mission. While he ate, a CIA supervisor went over the flight plan one last time and waited for the coded go-ahead message from Washington.

• • •

With the time difference, it was 6:00 PM on August 27 in the District of Columbia, and the afternoon rush hour was just beginning. Richard Bissell sat in his downtown office on H Street, across from the Metropolitan Club, and waited for Allen Dulles to call with the mission's final authorization. Spread out on his desk was a map of Soviet central Asia with the rough geographic bearings of the Tyura-Tam ICBM test site, the latest weather reports indicating clear skies over Kazakhstan, and a copy of Jones's flight plan.

Jones did not know that he worked for Bissell, though some U-2 pilots would later reveal that they had heard rumors that their orders came from a "Mister B in Washington." Nor did the lawyers and ordinary business executives who worked in the suites next to Bissell's office in the Matomic Building realize that their tall, avuncular neighbor with the round-rimmed glasses and easy smile ran the CIA's most ambitious and classified program. Such was the secrecy surrounding Bissell's operation that to maximize security, the U-2 reconnaissance program was housed separately from the agency's main headquarters near the Lincoln Memorial.

The U-2 had grown out of the same 1954 Killian report that had warned Eisenhower of Soviet missile gains and recommended that the United States fast-track its Atlas ICBM program to keep pace. "We must find ways," the report had also stated, "to increase the number of hard facts upon which our intelligence estimates are based . . . and to reduce the danger of gross overestimation or gross underestimation of the threat." There was a virtual information blackout on the Soviet Union, as the bomber gap would amply demonstrate, and new technologies had to be harnessed to collect more accurate data about Russian intentions. Specific recommendations were made in a separate, more classified report, which was circulated within a narrower audience at the National Security Council. It had been prepared by Edwin Land, a flamboyant Harvard dropout whose Jewish grandparents had emigrated from the very same part of Odessa where Korolev had grown up. Land had founded the Polaroid Company and was known for his "spellbinding performances" at Polaroid annual stockholders meetings, where he wooed investors like "a Broadway star."

The United States, Land urged in his corollary report, had to begin the immediate development of two different types of high-altitude photo-reconnaissance platforms. The first of these, the construction of a state-of-the-art surveillance plane, was approved by Eisenhower in late 1954; the second, a technologically more complex option that involved outer space, struck some at the time as the stuff of science fiction. It was put on the back burner.

At the CIA, the forty-five-year-old Bissell was put in charge of the spy plane project. A Groton and Yale man, and an amateur ornithologist with the right WASP connections, Bissell hailed from a prominent Connecticut family with interests in the insurance and railroad industries. He had grown up in the famous Mark Twain House—the rambling Victorian mansion that Samuel Clemens had built with his literary proceeds—and he had summered aboard his family's string of ever larger yachts, fostering a lifelong love of the sea. Trained as an economist, Bissell had presided over some of the financial aid programs disbursed in Europe after the war under the Marshall Plan, earning a reputation as a good planner with a disdain for convention and bureaucracy. Like Medaris, he possessed a rebellious nature when it came to following rules, and he was notorious for driving the wrong way down one-way streets in Washington whenever he was in a hurry.

At once charming and aloof, garrulous and yet secretive, Bissell was the sort of highborn gentleman scoundrel that the CIA, under Allen Dulles, loved to recruit. At the agency, he quickly proved his organizational mettle by helping to orchestrate the overthrow of the leftist government of Jacobo Arbenz in Guatemala, which had threatened the interests of the politically connected United Fruit Company. (Allen Dulles was a significant shareholder in the company, which John Foster Dulles had once represented as legal counsel. The director of the National Security Council, Robert Cutler, had sat on United Fruit's board, and Spruille Braden, the assistant secretary of state for Latin American affairs, and Walter Bedell Smith, the undersecretary of state, would both join United Fruit's board after the coup.)

Bissell's success in Guatemala, and a similarly staged restoration of the pro-Western shah in Iran (to preserve the holdings of the Anglo-Iranian Petroleum Company), had marked him as a rising star within the agency. Allen Dulles was said to be grooming Bissell as his successor, and the two shared a love of sailing and socializing, though Bissell was not blessed with his boss's famous wit or his infamously roving eye. (Dulles's marital indiscretions were perhaps the CIA's worst-kept secret.)

To find the right airplane for the job, a craft that could fly unmolested deep into Soviet territory, Bissell turned to Clarence "Kelly" Johnson, the legendary aeronautical engineer who headed Lockheed's most classified "black" military programs. Johnson had a blueprint for such a super-high-altitude jet-glider, the CL-282 Aquatone. He had proposed it to the air force, but Secretary Quarles had opted to go with a rival design by Bell Labs, the company he once headed, to reconfigure British bombers for reconnaissance duty. The Aquatone had been deemed too frail and ungainly for regular air force service. Besides offending the air force's aesthetic sense, the radical weight reductions that would allow it to fly so long and so high necessitated cutting too many standard safety and redundancy systems. But it suited Bissell's needs. He had only two requests: it had to be built quickly and quietly.

The contract was shrouded in such secrecy that the Aquatone was listed innocuously as "utility plane number two," hence its eventual designation as the U-2. Only eighty-one people at Lockheed had been permitted to work on the Aquatone/U-2, in contrast to the thousands who typically labored on such projects, and they finished the prototype in a record eighty-eight days. During that time, janitors were not even permitted in the hangar where it was assembled. To further minimize potential security leaks, Bissell demanded that subcontractors deliver component parts to front companies at fictitious addresses, and he bypassed regular accounting procedures by paying for the plane with a series of $1,256,000 checks made out personally to Johnson and hand-delivered to his Encino home address. As far as Lockheed and the U.S. government were concerned, the U-2 was entirely off the books. Even within the White House staff, only two people—Eisenhower's personal assistants General Andrew Goodpaster and Gordon Gray— initially knew of the plane's existence.

A similarly circuitous route had been used to recruit and train air force pilots, who were interviewed in dingy motels around the SAC bases in Georgia and Texas and sent for reconnaissance training to a nuclear testing ground near Groom Lake, Nevada, where Bissell reasoned that the fear of radiation poisoning ensured privacy. Formally engaged as civilian employees of the Second Provisional Weather Squadron, Bissell's boys operated under the cover of high-altitude weather research. Like the unmarked planes they flew, Jones and his fellow aviators carried no identification papers or dog tags, and no regimental crests or badges adorned their flight suits. Before each mission, their undergarments were carefully vetted to remove identifiable features that might point to a U.S. manufacturer. Even incriminating American accents could be rendered stateless in the event of capture with the one item supplied to all U-2 pilots in addition to the revolver, packets of rubles, French francs, and gold trinkets they carried in their zipper pockets: a glass cyanide capsule, the "L" suicide pill. "The ampoule should be crushed between the teeth. The user should then inhale through the mouth," the CIA manual instructed. "It is expected that there will be no pain, but there may be a feeling of constriction about the chest. Death will follow."

The gun and money were mostly for effect, to make the pilots feel better about their chances of survival, which Allen Dulles privately estimated at one in a million. "We told Eisenhower that it was most unlikely that a pilot would survive," Bissell recalled, "because the U-2 was a very light aircraft, more like a glider, and would disintegrate" if it were shot down. Only three bolts, for instance, connected the tail section to the fuselage. "Holy smokes, this thing is made out of toilet paper," the test pilot Bob Ericson had exclaimed on first seeing the craft. During a training exercise in Germany, a U-2 had broken apart by simply flying into the turbulent wake of another jet. If anything ever went wrong, Eisenhower was assured, there would be no evidence of the intrusions. The president, however, had not been entirely convinced. "Well, boys," he had said, "I believe the country needs this information, and I'm going to approve [the program]. But I'll tell you one thing," he added prophetically. "Some day one of these machines is going to be caught, and we are going to have a storm."

Bissell, Eisenhower later recalled, had agreed with his assessment of the political dangers. But John Foster Dulles, ever the hawk, "laughingly" scoffed at their concerns. "If the Soviets ever capture one of these planes, I'm sure they'll never admit to it," he said haughtily. "To do so would make it necessary for them to admit also that for these years we have been carrying on flights over their territory while they had been helpless to do anything about the matter."

Secretary Dulles's overconfidence, and his policy of purposefully provoking the Soviets, would later earn scorn from historians, who would label him "reckless." But on that day he carried the argument, as he often did. Eisenhower relied on his foreign policy adviser and trusted his judgment on international affairs, even if he had reservations about his frosty personality, as his diary entries made clear. Some said Ike was even a little intimidated or scared of Dulles, who was frequently so forceful with the mild-mannered president that Bissell initially wondered "who was really in charge" in the White House. But after observing the pair's interactions for several years, Bissell came to the conclusion that Eisenhower really ran the show.

• • •

Shortly before 5:00 AM in Lahore, E. K. Jones was "integrated" into a fully pressurized suit and began breathing pure oxygen for the hour prior to takeoff. He did this to lower the nitrogen level in his blood to avoid getting the bends, in much the same way deep-sea divers decompressed in special chambers before surfacing—only in reverse, as he would be climbing rather than descending. The pressurized suit Jones wore would keep his body from boiling and exploding at the altitude he would travel, a height where the air is so thin that atmospheric pressure drops to one-twenty-eighth that of sea level. Gases at such low-density atmospheres expand rapidly, and the boiling point of liquids falls to ninety-eight degrees, just below body temperature. Wernher von Braun had demonstrated this effect, known as Boyle's law, for television viewers on one of his Disney programs. Explaining why astronauts would need pressurized suits to survive, he had shown training footage of a special chamber designed to simulate the vacuum of outer space. Inside the chamber was a beaker of water at body temperature. At 14.7 pounds per square inch, the atmospheric pressure at sea level, the liquid was stable. But when a thin membrane sealing off a powerful suction device was popped, the air was instantly sucked out of the chamber, the pressure fell to near zero, and the water in the beaker suddenly convulsed with bubbles and boiled over. "This is what would happen to an astronaut's blood if he was not wearing a protective suit," von Braun explained somberly.

Jones would not be traveling into space. But at the height the U-2 reached, he would skirt the edge of the earth's atmosphere, and his special orange flight suit would keep the gases in his intestines from expanding and bursting like overinflated balloons if the U-2's cabin pressure suddenly failed or if he had to parachute out of the plane. Unfortunately, there were no further contingency plans for either of those unpleasant scenarios. To save weight, the early versions of the U-2 were not equipped with an ejection system or a burdensome second engine. The plane had to travel light to reach its lofty objectives, and redundancy systems would weigh it down. Nor had the designers bothered to install a long-range radio, because in case of a mechanical malfunction deep in enemy territory, no aid would be forthcoming. The U-2's only defense was altitude; no other aircraft could fly as high. As long as the plane stayed at its ceiling of 70,000 feet, it was untouchable. But if its single J57 turbo-jet engine malfunctioned, stalled, or flamed out, as it had a nasty habit of doing, and the pilot had to dip below 40,000 feet to restart it, he would be exposed and completely helpless. The men who flew U-2s understood that if something went wrong, there were no backup systems. They themselves were not expected to survive. In fact, they were under instructions not to survive, and already five had died during training exercises.

"I was assured that the young pilots undertaking these missions were doing so with their eyes wide open," Eisenhower later wrote, "motivated by a high degree of patriotism, a swashbuckling bravado, and certain material inducements."

It took a special breed of person to accept the U-2's peculiar conditions of employment. The high hazard pay—three times regular air force salary, plus bonuses—could not have been the principal motivation for volunteering into the airborne espionage service. To be sure, the money was nice. Some U-2 pilots drove Mercedes convertibles, and Jones earned more than twice the annual wage of a senior technocrat like von Braun, another Mercedes aficionado. But risk rather than reward was the primary recruiting tool. At the Lovelace Clinic near Albuquerque, where pilot prospects were sent, Bissell had devised a very elaborate psychological profile of the sort of men he sought: patriots, naturally, but people who liked living on the edge, for whom death was not a deterrent.

Bissell's pilots had to meet one other critical criterion: they had to be

exceptionally gifted airmen, because the U-2 was among the most difficult aircraft to fly. Simply positioning the plane for takeoff required
great skill. Its turning radius of 300 feet was nearly ten times that of a
regular fighter jet, and visibility from the cockpit was virtually nonexis-
tent. The plane had another troublesome characteristic. At high altitude, where even the U-2's 200-foot wingspan barely generated lift in
the thin air, its 505-miles-per-hour stall speed and 510-miles-per-hour
maximum speed converged in what pilots called the "coffin corner,"
leaving a scant margin of error. On landing, the massive wings—three
times longer than the sixty-foot plane itself—also required deft handling. To get the U-2 down safely, the pilot had to stall the plane precisely two feet off the tarmac, exactly on the center line, and keep the
massive wings off the ground by flying the aircraft down the runway in
perfect equilibrium. If the U-2 tilted a few feet to either side, a wing
tip could slam into the concrete and sheer off or send the plane cart-
wheeling. The balancing act was all the trickier as it had to be performed after nine grueling hours of unremitting stress—of working
without food or drink or respite from the tension of flying over enemy
territory.

At a little past 8:00 PM Washington time (6:00 AM in Lahore), Bissell's phone finally rang. It was Allen Dulles. He and his brother had just
spoken to Eisenhower. Every U-2 flight required presidential approval,
and the "brothers-opposite"—Allen, affable and attractive in his tweed
jacket and pipe; Foster, rigid and righteous in his somber suits—had
jointly persuaded Eisenhower that mission number 4058 of Operation
Soft Touch was a "Go."

Half an hour later, Jones was lined up at the edge of the Lahore runway, pointing his U-2 into the rising sun. Ground crews popped the
safety pins from the wheeled "pogo" outriggers at the end of each wing,
and the pilot gunned the big Pratt and Whitney engine. As the wings
gathered lift, the outriggers fell away, and Jones put the plane into a
steep incline of 15,000 feet per minute. It was at this moment that the
craft was most vulnerable to being photographed by KGB spies, and
Jones hastened to recede from view. But he had to be careful to taper
off his ascent after 35,000 feet. Boyle's law affected the expanding gases
in the fuel tanks in much the same way as it did the human body. A U-2
had exploded once when the pilot climbed too high too fast and his

tanks blew up. So Jones eased off the control pedals to reduce his rate of climb. Soon the U-2 was a speck in the Pakistani sky as it continued its ascent beyond the range of telephoto lenses. At 70,000 feet, as the outside temperature dropped to 160 degrees below zero, Jones leveled the U-2. The skies above blackened and filled with stars, and over the horizon Jones could see the blue and white curvature of the earth. Beneath him, the mountain passes of the Hindu Kush unfolded like an accordion; beyond that, Afghanistan, and the endless orange plains of Soviet central Asia. He pointed the plane north and crossed into Soviet airspace.

• • •

The first U-2 mission over the Soviet Union had coincided with a good-will visit by Nikita Khrushchev to Spaso House, the U.S. ambassador's official residence in Moscow, on July 4, 1956. While Khrushchev toasted America's 180th birthday with Ambassador Charles Bohlen, a U-2 snapped aerial photographs of the Kremlin before heading off to photograph the naval and air bases around Leningrad.

Allen Dulles had worried about the timing of the mission and "seemed somewhat startled and horrified to learn that the flight plan"—which had included a pass over Poznan, the scene of Polish rioting only a few days earlier—"had covered Moscow and Leningrad," Bissell recalled. "Do you think that was wise the first time?" Dulles asked.

"Allen," Bissell replied, "the first time is always the safest," since the Soviets were not expecting the mission. But he was wrong. Bissell had presumed that because the U-2 had evaded most American radars during its test, the Soviets would not be able to pick it up either. What he didn't realize was that the USSR had recently deployed a new generation of radar capable of tracking planes at much higher altitudes.

Khrushchev had immediately been informed of the flight and viewed the timing of the incursion as a personal affront. The way he saw it, the Americans had humiliatingly thumbed their noses at him, violating Soviet airspace even as he stood on U.S. sovereign diplomatic soil, and challenging him to do something about it. Worst of all, he had been powerless to respond. Soviet air defenses had nothing in their arsenal that could hit the U-2. MiG-19 and MiG-21 fighter jets buzzed like angry hornets under the U-2, catapulting themselves as high as possible, but their conventional engines and stubby wings couldn't generate the necessary lift to reach it. Antiaircraft batteries sent useless barrages that also fell well short. Only the new P-30 radar had been able to track the intruder with a surprising degree of sophistication, as the diplomatic protest the USSR privately filed on July 10, 1956, indicated.

> According to fully confirmed data, on July 4 of this year, at 8:18 AM, a twin-engine American medium bomber departed the American occupation zone in west Germany, flew over the territory of the German Democratic Republic and entered Soviet airspace at 9:35 AM in the area of Grodno from the Polish People's Republic. The plane violated the airspace of the Soviet Union, following a course which took it over Minsk, Vilnius, Kaunas, and Kaliningrad, penetrating up to 320 kilometers into Soviet territory and spending one hour and thirty-two minutes over it.

The accuracy with which Soviet radar had plotted the U-2's course had stunned Bissell, though he took some comfort in the fact that the Russians had mistakenly identified the spy plane as a twin-engine medium bomber. At least that meant that the U-2's secret was still safe, that the Soviets had no inkling of what they were dealing with, or how to counter it. In its protest, which was not made public, the Kremlin had purposely omitted mentioning the plane's detour over Moscow and Leningrad. That was simply too embarrassing to admit. "What I remembered most about the U-2 flight is how reluctant Father was to send a protest note to the U.S. government," Sergei Khrushchev recalled. "All his injured pride resisted. . . . He thought the Americans were chortling over our impotence."

For Khrushchev, this latest incursion had come on the heels of Curtis LeMay's mock attack on Siberia. Only this time, the Americans had not overflown some remote corner of Russia's empty Arctic wasteland. They had brazenly put a plane right over Red Square, the symbol of Soviet power, and the most heavily defended piece of airspace in the entire Communist bloc. It was a provocation designed to end any chance of rapprochement. "Certain reactionary circles in the United States," the Soviets protested, in a thinly veiled swipe at LeMay and the Dulles brothers, were trying to sabotage "the improvement of relations" between the two countries. The Soviets openly blamed "renegade" elements in the U.S. Air Force, though, as John Foster Dulles had predicted, they were careful to keep their complaints quiet.

Secretary Dulles responded, disingenuously, that no U.S. "military" plane had violated Soviet airspace on July 4. Technically this was true, since the U-2 was a CIA operation. But it was also true that American military aircraft had been probing Soviet air defenses ever since the end of the Second World War. The forays, or "ferret" missions as they were known, used a series of converted bombers—initially propeller-driven RB-29s, then the bigger jet-powered RB-47s—to search for gaps in Russia's radar coverage and to determine how quickly the Soviets could scramble interceptors in response. Invariably, American planes would only brush up against Soviet airspace, making quick dashes across the frontier. Usually "ferret" pilots skirted the twelve-mile offshore territorial limit claimed by Moscow. Very occasionally, as in the case of LeMay's Operation Home Run, they penetrated the deeper three-mile limit set by international law. And always they fled at the first sign of an answering plane. It was a cat-and-mouse game that could at times turn deadly. (The fate of 138 U.S. airmen shot down during the border overflights remains unknown to this day because Washington never inquired as to their whereabouts, and their families would not be told for forty years. "Representations and recommendations have been made to me by intelligence authorities," wrote one State Department official after a C-118 with nine crew members aboard was shot down over the Baltic on June 18, 1957, "that no legal action be pursued.")

The sheer volume of ferret missions, several thousand a year in the mid-1950s, annoyed the Kremlin. But as long as the American planes stayed close to the borders, and the White House stayed silent if any of its planes were hit, the game was played within the acceptable limits of superpower rivalry. Of course, Washington might have had a different view of permissible cold-war norms if the situation had been reversed, and Soviet planes patrolled the American coastline, buzzing over New York or Los Angeles. "It would have meant war," Khrushchev told his son.

The U-2, however, raised the intrusions to a different order of magnitude. With the maiden Independence Day flight, the United States had abandoned any pretense of respecting the territorial integrity of the Soviet Union. U-2s didn't take tentative steps along frontiers. They flew border to border in brazen 4,000-mile north–south sweeps. Khrushchev, naturally, was livid at the sudden change of rules. To add insult to injury, the CIA repeated the July 4 overflight the very next day and followed up with four more flights over the next six days. To the Soviets, the seemingly ceaseless parade of American planes over their two largest cities was a humiliating signal that the hard-line hawks in Eisenhower's administration now intended to harass the USSR on a weekly basis.

"The notion that we could overfly them at will must have been deeply unsettling," Bissell acknowledged. But the information the U-2s were bringing back was worth the risk, as a jubilant July 17 CIA memo indicated. "For the first time we are really able to say that we have an understanding of what was going on in the Soviet Union on July 4, 1956," wrote the analyst Herbert I. Miller.

> Broad coverage of the order of 400,000 square miles was obtained. Many new discoveries have come to light. Airfields previously unknown, army training bases previously unknown, industrial complexes of a size heretofore unsuspected were revealed. We know that even though innumerable radar signals were detected and recorded by the electronic system carried on the mission, fighter aircraft at the five most important bases covered were drawn up in orderly rows as if for formal inspection on parade. The medium jet bombers were also neatly aligned and not even dispersed to on-field dispersal areas. We know that the guns in the anti-aircraft batteries sighted were in a horizontal position rather than pointed upwards and "on the ready." We know that some harvests were being brought in, and that small truck gardens were being worked. These are but a few of the examples of the many things which tend to spell out the real intentions, objectives and qualities of the Soviet Union.

The "bomber gap," the reconnaissance flights soon showed, was bogus. There were no new armadas of Bears and Bisons lining Russian runway, just row after row of smaller shorter-range Tupolevs that could never reach American soil. But Ike couldn't confront Senator Symington with this information; it would mean blowing the U-2's cover.

Aside from the treasure trove of data it produced, the beauty of the U-2 lay in its deniability. As long as the Russians couldn't produce hard physical evidence of the incursions, or were too ashamed to make a public fuss, the planes could operate with impunity. As a result, as Miller's memo underscored, the CIA for the first time could eliminate much of the guesswork about Soviet weapons development programs and arms buildups. Craters at nuclear test sites could be photographed and measured to determine the size of the blasts. Missile launch sites could be examined for clues as to the capabilities of Russian rockets. Submarine pens could reveal the secrets of the Soviet underwater flotilla. Air base photographs could give an accurate picture of the number, strength, and battle readiness of bomber fleets. In a society so closed that it took six weeks for the CIA to get wind of Khrushchev's not-so-secret speech (despite the mass protests that it set off in Tbilisi), the best way to peer past the Iron Curtain was from above. A lone U-2 could produce infinitely more useful data than all the previous reconnaissance missions combined. What's more, the information could be targeted, aimed at a particular site the CIA wanted to know about. And on August 28, 1957, the highest-value target in the Soviet Union was Tyura-Tam.

• • •

If the R-7 was no longer a secret, it was partly Nikita Khrushchev's fault. He had been unable to resist trumpeting the achievement, thumbing his own nose at the Americans a little, and had ordered TASS, the official Soviet news agency, to issue a vague but suitably ominous announcement on August 26 heralding the triumphant test flight.

> A few days ago a super-long-range, intercontinental multistage ballistic missile was launched. The tests of the missile were successful; they fully confirmed the correctness of the calculations and the selected design. The flight of the missile took place at a very great, hereto unattained, altitude. Covering an enormous distance in a short time, the missile hit the assigned region. The results obtained show that there is a possibility of launching missiles into any region of the terrestrial globe. The solution of the problem of creating intercontinental ballistic missiles will make it possible to reach remote regions without resorting to strategic aviation, which at the present time is vulnerable to modern means of anti-aircraft defense.

After the humiliations of the U-2, Khrushchev had been only too happy to rattle his own saber for a change. But even if he had not wanted the Americans and the British to tremble at news of his new superweapon, the United States would have known about it anyway. The new National Security Agency, the sister organization to the CIA for signals intelligence gathering, had encircled the Soviet Union with an electronic moat. Huge dish and phased-array radar networks in Norway, Britain, Greenland, Germany, Turkey, Iran, Pakistan, Japan, and Alaska intercepted Soviet communications and tracked weapons tests. An American installation in northern Iran, Tacksman 1, had been monitoring the R-7 trials from the beginning and had been able to triangulate the general vicinity of the launchpad at Tyura-Tam.

From its dish network atop a 6,800-foot peak in Iran's Mashad mountains, the NSA had been able to follow the initial failures at Tyura-Tam on its radar screens, but the success of Korolev's fourth attempt had caused serious consternation in military circles. The R-7's maiden flight had coincided, almost to the day, with the fourth consecutive launch failure of the U.S. Air Force's intermediate-range Thor missile. The American equivalent of the R-7, the three-engine Atlas ICBM, was still a year away from a full flight test. To date, the Atlas had flown only under partial power, with only two of its three engines firing. That left Wernher von Braun's modified Redstone, a research rocket known as the Jupiter C (though it had nothing to do with the endangered Jupiter IRBM program), as the closest operational American response to the Soviets' ICBM breakthrough. But the Jupiter C, which had covered a 1,200-mile trajectory several weeks earlier during an August 8, 1957, trial, was not a missile, and hence it was not bound by the 200-mile limit imposed on the army by Charlie Wilson. Medaris had simply chosen its now ill-fated name before Wilson's "roles and missions" edict, as a way to gain access to overcrowded launch sites, which gave preference to military missiles over research projects, by making the test rocket sound as if it were part of the Jupiter program. In reality, it was solely a test vehicle with no deadly payload, designed to determine whether the new heat-resistant nose cone materials could withstand the pressures of high-speed atmospheric reentry.

Like the Soviets, American rocket scientists were grappling with the problem of warheads being incinerated on reentry, but, unlike Korolev, they had decided to test their thermal nose cone shield before perfecting the missiles that would carry the warhead. The Pentagon did not know that Korolev had put the cart before the horse, and that the Chief Designer did not yet have a working nose cone. The American military planners knew only that the Soviet Union claimed to have an operational ICBM, while the United States was still struggling to get a working IRBM.

This alarming imbalance was why Jones was flying over the Kazakh desert in search of Tyura-Tam. He was following the thin outlines of rail spurs, since the CIA believed that the Soviets could move their big missiles only on trains. And he was using old World War II German maps to guide him, since much of the Soviet landmass beyond the Urals was a mystery to American cartographers. His was only the fourteenth U-2 overflight into the Soviet Union. Eisenhower had twice ordered the flights stopped: first after the Kremlin had filed its initial diplomatic protest, and then again after the sixth mission in November 1956, when the pilot Francis Gary Powers had experienced electrical problems over the Caucasus with MiGs hot on his tail. The scare had sobered some U-2 enthusiasts in Washington, who feared an international incident. Despite the wealth of intelligence gleaned from the U-2s, the superpower tensions they provoked frightened Eisenhower. Put simply, they were driving Khrushchev insane with anger. "Stop sending intruders into our air space," the Soviet leader had railed at a stunned delegation of visiting U.S. Air Force generals in the summer of 1956. "We will shoot down uninvited guests. . . . They are flying coffins."

"At that moment," recalled a Soviet participant, Lieutenant Colonel Alexander Orlov, "Khrushchev noticed that a U.S. military attaché was pouring the contents of his glass under a bush. Turning to U.S. Ambassador Charles Bohlen, the Soviet leader said, 'Here I am speaking about peace and friendship, but what does your attaché do?' The attaché was then pressured into demonstratively drinking an enormous penalty toast, after which he quickly departed."

Juvenile drinking antics aside, Khrushchev had been dead serious about shooting down American intruders and closing the technology gap that allowed the U-2 to range over Soviet soil with such arrogance. "Father thirsted for revenge," Sergei Khrushchev recalled. Immediately, he had ordered that development of the SU-9 supersonic high-altitude fighter jet be fast-tracked, and Soviet rocket scientists were given top priority to push ahead with the new SA-2 surface-to-air missile, which would raise the strike ceiling from 50,000 feet to 82,000 feet.

The CIA knew that it was only a matter of time before the American height advantage was lost. But after Moscow's brutal suppression of the Hungarian uprising, the Dulles brothers pushed Eisenhower to resume the suspended flights. They felt guilty about not having come to Hungary's aid and wanted, perhaps, to lash out at Khrushchev—the "Butcher of Budapest," as Vice President Nixon had taken to calling him. On the secretary of state's advice Nixon had been dispatched to the teeming refugee camps in Austria, where over one hundred thousand Hungarians wallowed in resentment. "They blamed us for first encouraging them to revolt, and then sitting back while the Soviets cut them down," the vice president reported on his return. That had settled it. If the flights irked the Butcher, so much the better.

• • •

The grainy black-and-white photographs of Tyura-Tam taken by E. K. Jones on August 28 were flown to Washington the following day. There, in another nondescript office that Bissell had rented—this one above a Ford repair shop at Fifth and K streets—a team of optical experts with large magnifying glasses pored over the 12,000 feet of negatives. The Tyura-Tam shots showed a deep triangular fire pit that looked like a large terraced rock quarry. This excavation absorbed some of the R-7's considerable exhaust blast at liftoff, and its sheer magnitude had given the CIA analysts pause. Perched over the pit was the near football-field-sized launch table with its towering Tulip jaws. Once again, the proportions appeared staggering compared to American launch stands of the period. From the Tulip, a gigantic berm with wide-gauge rail tracks ran toward an imposing assembly building about a mile away. The CIA could have drawn only one conclusion from the images: the R-7 was a monster.

Once more, the U-2 had proved its worth. Only this time it had also proved Stuart Symington right. Having exposed the "bomber gap," the Missouri senator had turned his attentions (and his presidential ambitions) increasingly toward rocketry. In early 1957 he had begun advocating increases in missile outlays, warning about Soviet advances. "I don't 'believe' that the Soviets are ahead," he said in a February appearance on NBC's *Meet the Press*. "I state that they *are* ahead of us."

Nonsense, scoffed his fellow panelist Donald Quarles, puffing up his small frame in a swell of indignation to challenge the tall and imposing senator. The United States "is probably well ahead of Russia in the guided missile race," he countered, with equal confidence.

The argument had spilled over onto the pages of national newspapers, where Symington accused the Eisenhower administration of misleading the American people. Threatening that Congress would launch a "searching inquiry," he asserted that the air force's missile budget needed to be doubled to keep pace with Russia. "Every day we don't reverse our policy is a bad day for the Free World," he thundered. By late summer, when Charlie Wilson announced his $200 million missile cuts, Symington had emerged as the administration's leading critic on matters of national security, the Democratic Party's preeminent cold warrior. Eisenhower, he groused, was engaging in "unilateral disarmament," endangering the nation with his obsession for balanced budgets. The Soviets were going to overtake America, he warned, and then all those new highways Ike was pouring billions into would serve only as evacuation routes.

Unbeknownst to Symington, he had a covert ally in Richard Bissell, who understood that it was only a matter of time before the U-2's luck ran out. The intelligence data it produced were invaluable, but the political costs were simply too high. Sooner or later a less risky means of collecting intelligence would need to be found. And Bissell already knew what that was.

The same 1954 Land report that had urged the creation of the U-2 had also made a recommendation for the development of another type of high-altitude reconnaissance craft, a satellite. The idea, at the time, had been met with skepticism by the National Security Council, owing to its technological complexity, though it was hardly revolutionary.

The notion of using the cosmos as a surveillance platform had long stirred the imagination of rocket scientists and spies on both sides of the cold war divide. As early as 1946, a West Coast military think tank, the RAND Corporation, had envisioned successors of von Braun's V-2 rockets one day carrying cameras beyond the stratosphere. Von Braun himself had made a similar pitch to the army brass in 1954. "Gone was the folksy fellow with rolled-up sleeves and Disneyesque props," wrote the historian William Burrows of the meeting. "He was replaced by a grim-faced individual with a dark suit who puffed on cigarettes from behind a desk. This von Braun explained that a satellite in polar orbit would pass over every place on Earth every twenty-four hours, a perfect route for robotic espionage. He noted that maps of Eurasia were five hundred yards off, and added that the error could be reduced to twenty-five yards. The implication was wasted on no one in the room: taking photographs of Earth from space not only would create an intelligence bonanza but would vastly improve targeting accuracy."

Neither von Braun's pitch nor Land's recommendation to the NSC had received much traction in 1954 because American rockets were still too small and underpowered to contemplate sending up heavy spy satellites. By 1957, however, missile development had progressed sufficiently that the notion no longer seemed far-fetched. Though Bissell was working on a successor to the U-2—a new plane made entirely of titanium that could fly at 80,000 feet at a speed of 2,600 miles per hour, nearly five times faster than the U-2—he was also thinking that satellites might offer a simpler long-term solution. The SR-71 Blackbird supersonic spy jet that he was developing with Kelly Johnson might prolong America's ability to sneak into Soviet airspace, but its invulnerability would also be only temporary. Eventually the Soviets would find a way of bringing it down, too. A satellite, on the other hand, could not be shot down.

Bissell had a problem, though. The CIA was not in charge of the satellite mission; the air force was—but the project was languishing on the shelf. Bissell was alarmed that it was not even at the blueprint stage. Worse, it apparently had received such a low security classification that articles were appearing on it in the aviation press—a death knell to any covert program. Compared to the secrecy that surrounded the U-2, it seemed to Bissell as if the air force was advertising its lack of interest in spy satellites, and in the process blowing the cover of what could potentially prove to be the intelligence community's premier surveillance tool.

The same lackadaisical attitude plagued the navy's quasi-civilian entry into the International Geophysical Year's scientific satellite competition. Vanguard, as the project was known, had been chosen by Quarles over von Braun's army proposal, sparking a furious rearguard campaign by Medaris to have the decision reversed. The navy's effort was mired in technical difficulties, hopelessly underfunded, and badly behind schedule. The four-stage Jupiter C test rocket, on the other hand, was already reaching near-orbital velocity in its trials and could easily escape gravity if it were allowed to use a live fourth stage, instead of the dummy weighted with sand that the Pentagon ordered.

Alas, after the Nickerson scandal, ABMA had few friends in the administration. Not only were Medaris's pleas gruffly rebuffed, but Engine Charlie spitefully ordered the general to personally inspect every Jupiter C launch to make sure the uppermost stage was a dud so that von Braun did not launch a satellite "by accident."

With the army sidelined, the navy bogged down with crippling delays, and the air force generally uninterested, Bissell was in despair. "I knew our national effort to put any kind of a satellite into orbit was lagging badly," he recalled. "Given my keen personal desire to implement a successor to the U-2 program, I approached Allen Dulles and urged him to take action. He gave me permission to meet with Deputy Secretary of Defense Donald Quarles to inform him that the CIA was interested in accelerating the development of a satellite. Perhaps threatened by my approach, the Pentagon added a modest sum of money to the Navy's budget to speed up its work, on the grounds that its satellite was the most promising of the candidates for an early flight. Unfortunately, the additional funds accomplished little."

There was another reason why Bissell began pushing Quarles, who had just been promoted from assistant secretary to deputy secretary of defense, to start taking the IGY satellite effort more seriously. A legal question was nagging CIA attorneys, one that threatened to scuttle the future of any orbiting surveillance system: Who had the territorial rights to outer space? Did sovereign airspace extend beyond the stratosphere? There was no legal precedent for a satellite circumnavigating the globe, snapping photographs of foreign countries. Would it violate international law, like the U-2?

The sooner a satellite was sent into orbit, the quicker a precedent would be set that would govern the legality of all future launches. In that regard, a purely scientific satellite, such as the Naval Research Center's entry into the civilian IGY competition, was the perfect foil for establishing the open, international nature of outer space that would make extraterrestrial spying lawful.

Legal issues aside, there was also the question of national prestige. One of the CIA's principal tasks was to engage the Soviet Union in psychological warfare, and Bissell worried that if the Communists were first in space, they would score a significant victory over the capitalist democracies of the West in the battle for the hearts and minds of the developing, postcolonial Third World.

Bissell was not the only one concerned about the propaganda value of a Soviet space milestone. The nuclear physicist I. I. Rabi, a future Nobel laureate, wrote to Eisenhower pleading for greater resources for the American IGY satellite "in view of the competition we might face" from Soviet science. "It was unfortunate," recalled James Killian, the author of the initial capabilities report recommending increasing both missile and reconnaissance spending, "that this advice did not produce any significant undertakings by Eisenhower."

Wernher von Braun, for his part, bypassed the reluctant administration altogether, taking his calls for greater action directly to what he thought might be a more sympathetic audience: the Democratic Congress. Employing the same tactic that Korolev had used to such effect on Khrushchev, he warned a Senate subcommittee that the Soviet Union was in the advanced stages of developing a satellite. But far from taking the bait, Senator Allen Ellender of Louisiana almost burst out laughing. "Ellender said that we must be out of our minds," recalled General James M. Gavin. "He had just come from a visit to the Soviet Union, and after seeing the ancient automobiles, and very few of them on the streets, was convinced we were entirely wrong."

"The Soviets," the senator scoffed, "couldn't possibly launch a satellite."

And yet, after the R-7's successful test flight, the official Soviet press became uncharacteristically voluble on the subject of satellites. Sergei Korolev himself made a rare public appearance on September 17 to celebrate the one hundredth anniversary of the birth of Konstantin Tsiolkowsky, Russia's first rocket scientist and space visionary. Speaking in the ornate Hall of Columns in Moscow, Korolev told a packed house of the Soviet Union's most senior academics of the R-7's triumph and promised that "in the nearest future, the USSR will send a satellite into space." The speech was reprinted in the following day's edition of *Pravda* under the pseudonym S. Sergiev. *Pravda* also ran a story by Dr. A. N. Nesmeyanov, the president of the USSR Academy of Sciences, which boasted, "The creation and launching of the Soviet artificial satellite for scientific purposes during the International Geophysical Year will play an exceptional role in unifying the efforts of scientists of various countries in the struggle to conquer the forces of nature."

Radio magazine in Moscow went a step farther, providing detailed instructions for amateur radio enthusiasts to receive the frequencies on which the future Soviet satellite would broadcast. Another trade publication, *Astronomer's Circular,* advised its readers, "The Astronomical Council of the USSR Academy of Sciences requests all astronomical organizations, all astronomers of the Soviet Union, and all members of the All-Union Astronomical and Geodetic Society to participate actively in preparations for the visual observation of artificial satellites."

The message from Moscow was loud and clear: a Soviet satellite would soon be orbiting the earth, and ordinary citizens would be able to see and hear it.

The warning signals did not fall completely on deaf ears in America. The *New York Times* started researching a story about an impending Soviet launch. The RAND Corporation also carefully clipped all the Soviet press briefs and forwarded them to the Pentagon with an appended note concluding that the Soviets must be serious. But in Washington, no one had time for talk of satellites. The country was in the throes of a looming crisis that had begun with the opening of the school year and was quickly escalating into a major challenge to President Eisenhower's authority.

• • •

Like Khrushchev during the R-7's string of failures, Dwight Eisenhower was preoccupied in the waning weeks of September 1957 by unfolding events that had nothing to do with rockets or the conquest of space, and everything to do with cleaning up a political mess that was largely of his own making.

The trouble had started with the U.S. Supreme Court's 1954 ruling in *Brown v. Board of Education*, which invalidated the "separate but equal" racial guidelines that had governed segregated schools. The decision had sparked a political rebellion in the heavily Democratic South, where nineteen senators and seventy-seven congressmen issued a defiant proclamation in March 1956, condemning the Court and its ruling. Only Albert Gore Sr. of Tennessee and Lyndon Baines Johnson, the majority leader from Texas, had refused to add their signatures to the "Southern Manifesto."

At the White House, Vice President Nixon had reacted furiously to this paean to segregation, advocating a strong response. There were not only constitutional issues at stake, but moral implications as well, and Nixon believed the administration could not stand idly by while the South thumbed its nose at the Supreme Court. Ike's inner circle, though, had different concerns. The 1956 election was just around the corner, and Eisenhower had been making considerable inroads with voters in the South. The time was not right to rock the boat. Already Nixon's more liberal stance on civil rights was causing problems in the South, and while Ike decided in the end not to drop him from the ticket, tension remained between the president and the vice president. The Democrats focused their ire on Nixon rather than on the popular Eisenhower. "It was hard not to feel that I was being set up," Nixon later reflected, noting his "disillusionment with the way Eisenhower was handling the affair."

The election confirmed both Eisenhower's high standing with the electorate and Nixon's lowly position in the administration's hierarchy. Ike clobbered Stevenson, but the Democrats easily carried both Houses, a sign that the vote had been more about Ike's personality than a partisan endorsement of the Republicans. Following the landslide victory, Eisenhower sent Nixon a belated note that must have only added to the vice president's growing sense of resentment. "Dear Dick," the president wrote in late December, well over a month after the election. "I find that while I have thanked what seems to be thousands of people from Maine to California for their help in the political campaign, I have never expressed my appreciation to you."

The new year brought renewed calls for action on civil rights, as African-American leaders such as Martin Luther King Jr. developed larger followings. Nixon lobbied hard for the White House to introduce legislation that would update the nation's ineffectual civil rights laws and in the process drive a wedge between the Democratic Party's more liberal northern wing and the reactionary factions in the South. But once more Eisenhower wavered, thinking the time not right, and the initiative languished. Into this void stepped Lyndon Baines Johnson.

Like Symington, Johnson was maneuvering for the 1960 presidential nomination and searching for politically promising causes. As majority leader, LBJ outranked Symington and was a far more seasoned and skilled legislator. But while the media hailed Symington as patrician and formidable, Johnson was dismissed as preening and pompous. Symington was the blue-blooded Yale man; Johnson, the graduate of a provincial Texas teacher's college with the uncouth manners of a shady oil tycoon. "He flashes gold cuff links, fiddles with the gold band of his gold wristwatch, toys with a tiny gold pill box, tinkers with a gold desk ornament," *Time* magazine noted with evident distaste. "His LBJ brand appears everywhere, on his shirts, his handkerchiefs, his personal jewelry, in his wife's initials, in his daughters' initials. . . . Lyndon Johnson would rather be caught dead than in a suit costing less than $200."

Symington was portrayed by the press as polished and statesmanlike, a leader immersed in the grave concerns of national security. Johnson, on the other hand, served hamburger patties shaped like the state of Texas at his Johnson City ranch, urging guests to "eat the panhandle first." Hardly presidential material.

Despite his outlandish public image, Johnson's position on civil rights was relatively moderate. And it was through this violently divisive issue that he saw a chance to make his mark.

The 1957 Civil Rights Act that he almost single-handedly wheeled through a reluctant Congress was at once a testament to his immense talents as a backroom negotiator and one of the most cynical compromises in modern American politics. "I'm going to have to bring up the nigger bill again," Johnson would privately apologize to southern senators, all the while pushing and cajoling them to give ground. They did, but they also extracted many concessions that watered down the spirit and letter of the proposed law. The final version served only to encourage southern Democrats to redouble their efforts to fight integration.

Throughout the process Eisenhower had sat conspicuously silent. On the few occasions when he weighed in on race relations, his statements were sufficiently contradictory and middle-of-the-road that each side simply chose to hear what it wanted to hear. Privately, he expressed his feelings more clearly. "Southern whites," he told Chief Justice Earl Warren, "are not bad people. All they are concerned about is to see that their sweet little girls are not required to sit in school with some big overgrown Negro."

Many southern Democrats saw Eisenhower's hedging as further license to defy the Supreme Court. "What he had not done was provide leadership, either moral or political," remarked the Eisenhower biographer Stephen Ambrose. "What he wanted—for the problem to go away—he could not have."

On September 4, 1957, the problem exploded. In Little Rock, Arkansas, nine black students attempted to register for classes at the all-white Central High School. A seething white mob forced them to retreat and nearly lynched one of the nine, fourteen-year-old Elizabeth Eckford. Governor Orval E. Faubus sided with the mob and called out the National Guard to prevent the black teens from registering.

Ike refused to dignify the situation by getting involved and made a determined show of sticking to his routine by going forward with a scheduled vacation to Newport, Rhode Island. Two weeks later, Eisenhower was still playing golf in Newport, trying to maintain a public facade of leisurely unconcern, while Governor Faubus brazenly continued to defy the highest law of the land. Finally, on September 20, the president summoned the rebellious governor to Newport to plead with him to follow a federal court injunction not to interfere with Little Rock's integration. "I got the impression at the time," Faubus later said of the meeting, "that he was attempting to recall just what he was supposed to say to me, as if he were trying to remember instructions on a subject on which he was not completely assured in his own mind."

Faubus made a vague, and as it turned out short-lived, promise to stand down, and on September 23, the nine black students were spirited into Central High School under police guard through a side door, as the crowd outside chanted, *"Two, four, six, eight, we ain't gonna integrate."*

With tensions rising, and the threat of violence increasing, the nine black students were summoned to the principal's office. A few were quietly crying but sat numbly, tugging nervously at the plaid skirts and pressed trousers their parents had purchased for the new school year. They were all excellent students, chosen by the NAACP to break Little Rock's color barrier because of their good grades and character. They watched as worried-looking officials streamed in and out of the principal's office. From down one of the halls came the sound of glass crashing.

From inside the principal's office, the students could hear the alarmed officials through the partially opened door. "We're trapped," said one frantic voice. "Good Lord, you're right," said another. "We may have to let them have one of the kids so we can distract them long enough to get the others out."

"*Let one of those kids hang?*" shouted another voice. "How's that gonna look? Niggers or not, they're children, and we got a job to do."

Soon, the door to the principal's office was flung open, and a tall, dark-haired man addressed the huddled teenagers. "I'm Gene Smith, Assistant Chief of the Little Rock Police Department," he said in a calm, kind tone. His had been the voice urging that all the children be saved. "It's time for you to leave today," he announced, leading them to an underground garage. "Come with me."

Little Rock's experiment with school integration had lasted less than three hours. "The colored children [were] removed to their homes for safety purposes," Mayor Woodrow W. Mann informed Eisenhower in an urgent telegram. "The mob that gathered was no spontaneous assembly. It was agitated, aroused, and assembled by a concerted plan . . . [which] leads to the inevitable conclusion that Governor Faubus was cognizant of what was going to take place."

Eisenhower was furious at Faubus's duplicitous double-cross. As the journalist David Halberstam later observed, "A man who had been a five-star general did not look kindly on frontal challenges by junior officers. After vacillating for so long, he came down hard, seeing the issue not as a question of integration as much as one of insurrection."

That same afternoon, the president ordered one thousand paratroopers from the 101st Airborne Division to Little Rock. It was the first time since Reconstruction that federal troops were deployed in the South. "Troops not to enforce integration but to prevent opposition to an order of a court," Eisenhower noted to himself on his personal stationery, along with doodles of an airplane, sundry checkmarks, scribbles, and several illegible musings.

By early October, the situation had been brought under control. But the resolution came at a high price for Eisenhower. "A weak President who fiddled along ineffectually until a personal affront drives him to unexpectedly drastic action" was how the former secretary of state Dean Acheson described Eisenhower in a letter to Harry Truman. "A Little Rock with Moscow," Acheson added, "and SAC in the place of the paratroopers could blow us all apart."

The Democrats, like Acheson, saw that Little Rock had badly dented Eisenhower's seemingly invulnerable image as a strong and decisive leader. What they didn't realize was that an even greater crisis of confidence loomed just days away. And as Acheson had feared, this time it would be the Russians who would test Eisenhower's mettle.

7

A SIMPLE SATELLITE

There was only one problem with Sergei Korolev's satellite plan. The R-7 State Commission, the government body that oversaw missile testing, wasn't buying it.

One after another, Korolev fixed the commission members with long, livid looks. *Fools*, his eyes blazed, *you stupid, shortsighted fools*. Thirteen hard and hostile faces returned his angry gaze. From behind the thick stacks of telemetry readings and mission reports in front of each representative, a few smug and barely concealed smirks swirled amid the cigarette smoke and steaming glasses of sweet tea. For once, their satisfied expressions seemed to say, the arrogant Chief Designer wouldn't get his way. This time, he wasn't going to steamroll over anyone.

The meeting had not gone as Korolev had hoped. It had started cordially enough, with a postmortem of the successful August 21 launch. Glushko, Pilyugin, Kuznetsov, and the other bureau heads had delivered reports on how their engines, valves, and guidance and gyroscopic systems had performed during the maiden flight, and all had been found satisfactory. The critical failure of the warhead shield on reentry had been discussed without any of the rancor that had plagued the disastrous summer trials, and recommendations had been made for further investigations ahead of the next scheduled attempt on September 7. All the heartache, health problems, and bad blood seemed to have been forgotten, now that the rocket had worked, and everyone basked in its collective glory. Glushko and Korolev had even exchanged supportive glances as the list of systems successes had been read off. It was only when the Chief Designer rose to speak that the convivial atmosphere abruptly changed. "I suggest we begin preparations to launch the artificial satellite," he said to stunned silence. "I mean to use the primitive satellite PS-1," he added, referring to the Russian acronym for Simple Satellite Number One, the scaled-down substitute for the original 2,700-pound Object OD-1 that the Soviet Union had intended to enter into the IGY competition before discovering that Glushko's engines wouldn't be able to lift such a heavy payload. The smaller satellite, Korolev went on, was ready and could be launched in October with minimal alterations.

All at once, the room burst into a chorus of competing protests. In the hue and cry, the smiles vanished, Glushko suddenly sneered, and voices were raised. "This proposal was a big surprise," General K. V. Gerchik, the Tyura-Tam deputy base commander, recalled decades later. "There were objections."

Rather heated objections, according to the sparse historical record, which, in typical Soviet fashion, whitewashed the unseemly bickering in bland bureaucratic terms. Apparently Korolev's fellow commission members had seen through his transparent ploy to distract Khrushchev from the warhead reentry problems by orbiting a satellite instead, and they wanted no part of the scheme. For their own reasons, the different factions on the commission rallied against Korolev's rush to orbit. There wasn't time, Glushko and the other design bureau heads objected, reminding the Chief Designer that the string of test failures had depleted the available stock of R-7s. They simply didn't have the component parts to build more rockets to keep pace with Korolev's frenetic schedule.

The six military representatives on the commission complained that a satellite was a waste of time and resources. The R-7 was a weapon, not a toy for silly scientific competitions. "All these space projects will simply distract us from the main objective of a nuclear ICBM," said General Aleksander Mrykin, reiterating the military's long-held position. "We should delay the development of a satellite until the R-7 is fully operational."

Major General Oleg Shishkin, the nuclear ordnance chief, was particularly irate. It was his dummy warheads that were being incinerated by the faulty thermal shields, and until Korolev fixed the problem he couldn't risk testing the R-7 with a live weapon. General Ivan Bulychev, the deputy communications commander, had equally pressing and practical reasons to oppose the satellite. It was Bulychev's ground stations that would need to be reconfigured to track a completely new orbital trajectory, and the impatient Chief Designer wanted the upgrades ready by early October—an unrealistic time frame. Colonel Yuri Mozzhorin recalled, "The Directorate of Missile Weapons was sharply against the participation of the Ministry of Defense in the tracking of satellites . . . because it would harm the defensive capabilities of the country."

Korolev had anticipated such rumblings from the armed forces. Soldiers were inherently conservative, and every major modern military innovation from the Gatling gun to the aircraft had been viewed with extreme skepticism by general staffs. Khrushchev had foisted missiles on his reluctant generals, had taken billions of rubles from their budgets to finance his gamble, and now the scheming Korolev was trying to bamboozle them with this sudden urgency to orbit a man-made moon. To deflect the criticism, the Chief Designer had already petitioned the government to issue a decree ordering his subordinate design bureaus to begin "development of an artificial satellite for photographing the earth's surface." Like von Braun three years earlier, he had hoped that the potential military applications of spacecraft as targeting and espionage vehicles would generate enthusiasm.

But the military men were unmoved. The Soviet Union did not have the same urgent need for aerial reconnaissance as the United States. America was a far more open society, ridiculously easy to penetrate. Its Congress publicly debated minute details of defense budgets. Astonishingly accurate road maps that would be highly classified in the USSR were sold at every gas station. Security was so incredibly lax that Korolev had been able to read a translation of U.S. newspaper accounts of Atlas's latest test, a fiery failure off the Florida coast, and of the successful recovery of a nose cone Wernher von Braun had blasted into space with his Jupiter C. Von Braun, conversely, did not even know that Korolev existed—a fact that apparently grated on the egotistical Chief

Designer. For the Soviets, it was mind-boggling how much information the Americans naively left lying around for the KGB to scoop up. Russia's generals didn't need a satellite to find out what was going on in Washington. They needed a missile that could destroy it.

Korolev's satellite surprise had also encountered resistance from another, entirely unexpected quarter: from Mstislav Keldysh, a legend in the ranks of Soviet academia, who had been awarded three Hero of Soviet Labor gold stars, the USSR's equivalent of the Congressional Medal of Honor. Korolev's junior by four years, Keldysh was a handsome Latvian mathematical genius, a child prodigy both as a theoretician and as an applied aviation engineer, who had been the youngest member ever elected to the prestigious Academy of Sciences. In Soviet scholarly circles, Keldysh's name carried the same awe-inspiring weight as that of Albert Einstein or Enrico Fermi. The suave Balt, a master at bureaucratic maneuvering, had used his status and unique position as the Presidium's science adviser to become the USSR's earliest advocate of space exploration. Long before Korolev's breakthrough with the R-5 intermediate-range missile, it had been Keldysh's political clout that had nudged the satellite proposal gradually up the government ladder.

But now the man who had once been Korolev's greatest proponent was openly skeptical. This flimsy "simple satellite" that Korolev wanted to launch was nothing more than a political stunt, Keldysh complained. They should wait until Object OD-1, the mammoth orbiting laboratory bristling with sophisticated sensors, was ready. That way the Soviet Union could score a real scientific coup. Object OD-1 would perform invaluable research such as measuring the density, pressure, and ion composition of the atmosphere at altitudes from 200 to 500 kilometers (124 to 310 miles). It would measure magnetic fields, cosmic rays, and the corpuscular radiation of the sun. Ultraviolet and X-ray spectrums, inherent electric charges, and positive ion concentrations could be determined—in short, everything scientists needed to know before attempting to send a living being into space. The PS-1 Simple Satellite that Korolev now proposed launching instead was so small that the only real task it could perform was to send short bursts of a low-frequency signal back to amateur ham radio operators on earth. As a research tool, it had no real value.

Keldysh now charged that Korolev was more interested in the personal prestige of breaking the space barrier than in the collective data the mission would bring back. This was all about the Chief Designer's ego, not the advancement of science, and he would be damned if he was going to tie up the Soviet Union's most powerful computer for Korolev's little vanity project. Russia had virtually no computers in 1957 because Stalin had viewed cybernetics as a "faulty science," not applicable to a dialectical society. By the time the military applications of the machines had become obvious, the USSR lagged hopelessly behind the West, and Keldysh, as the head of the Steklov Institute of Applied Mathematics, controlled access to the only civilian supercomputer in Moscow.

Without access to Keldysh's computer, Korolev would be stopped in his tracks because it could take months using manual six-digit trigonometry tables just to plug in all the variables needed to plot the parameters of an orbital trajectory. His engineers needed to calculate the exact speed of the rotation of the earth at the point of launch, which, at Tyura-Tam, was just over 1,000 feet per second. The direction of the launch, the azimuth, had to be factored in, since the earth rotates on a west-to-east axis and launching westward would be like swimming against a tide. The precise shape of the earth at the point of launch also had to be measured, since the planet is not a perfect sphere. The inclination of the equatorial plane, the angle between the equator and the azimuth, then needed to be calibrated to determine fuel load, which in turn affected the mass-to-thrust ratio critical to calculating the "escape velocity" that would propel PS-1 beyond the pull of gravity. This in turn determined the apogee and perigee—that is, the peaks and troughs of the satellite's wavelike orbit—along with the duration of each revolution. These variables were all mercilessly interrelated, and the smallest mistake could result in the satellite crashing back to earth or escaping into deep space, never to be seen again. Even if PS-1 were catapulted to its proper celestial position, a tiny error in trajectory computations could result in a widely errant elliptical orbit that would either bypass the United States or appear over the North American continent at the wrong times. That, for Korolev, would spell disaster, because he wanted his little satellite seen in the night sky over enemy territory. It was why he had ordered PS-1 made entirely of a highly reflective aluminum material, polished to a mirrorlike sheen, and why he had gone to such lengths to insist on a spherical shape. Korolev had shot down the cone, the cylinder, the square, and every design his frustrated satellite makers had proposed. "Why, Sergei Pavlovich?" one of them finally asked, exasperated. "Because it's not round," he had replied mysteriously.

There was no real mystery, however. Spinning spherical objects simply caught the light better, and PS-1 would act like a bright mirror for the sun's rays as it circled the earth, making it much more visible in the dark. Without this form of optical amplification, PS-1, at twenty-two inches in diameter, was too small to be seen from distances of up to 500 miles away. And seeing, as the old saying went, was believing.

The same psychological reasoning applied to Korolev's decision to sacrifice scientific instrumentation in favor of audio capability. Virtually all of PS-1's 184-pound mass was consumed by two transmitters and their three batteries. The silver-zinc chargers alone weighed 122 pounds, providing power for only a few weeks of operation. As a redundancy, they operated two identical one-watt transmitters that broadcast alternatively on different frequencies using separate pairs of ten- and eight-foot antennae. This way if one system failed, a signal would still reach earth. Hearing was also believing.

Sights and sounds from space would give even a crude little craft like PS-1 enormous propaganda and political value, Korolev argued. No one would be able to deny its existence, and even the "simple satellite" would be a Soviet triumph over the Americans, orbiting proof of the supremacy of Communist countries. "The Soviet Union must be first," he said adamantly. Korolev's colleagues, though, were far from convinced. "The Army needs just one thing," Marshal Mitrofan Nedelin shot back, "a rocket that will work." The standoff continued, with Korolev demanding that the commission accept his proposal.

Glushko and the military men led the revolt—the Red Army representatives because it was a distraction from solving the nose cone problem, Glushko perhaps as a way of getting even with his rival. Whichever the case, all the old animosity had bubbled to the surface: the fights, "using the dirtiest language and crudest phrases," the tirades that had left so many wounded feelings. "Mindless malice," Korolev complained to Nina.

Smarting from the rejection, Korolev stormed out of the conference room, no doubt leaving his own habitual trail of expletives that Soviet historians opted not to record for posterity. He was back two weeks later, early in the second week of September, once again insisting that his Simple Satellite Number One be given the green light. This time, however, his position had improved, and he wasn't going to take no for an answer. The R-7 had completed another nearly flawless test flight on September 7, using up the last of the original batch of rockets, and except for the persistent nose cone meltdowns he now had back-to-back successes to offset the three failures and months of catastrophic delays. Parts for another rocket were already being shipped to Tyura-Tam from Moscow, so he had the hardware. Once more, Korolev also had Khrushchev firmly in his corner, and his benefactor, after outwitting the hard-line coup plotters, now sat alone atop the Soviet hierarchy as the undisputed master of the Kremlin.

For Khrushchev, the R-7's second consecutive success had also been a vindication of his vision for a new defense shield against the forces of imperialism. If the Soviet leader harbored any regrets with regard to the R-7, it was the disappointing reaction in the West to his announcement of the new weapons system. Washington had not quaked in panic at news of the Communist ICBM, as Khrushchev had hoped. In fact, the American general public barely noticed, and the Eisenhower administration appeared dismissively skeptical as to whether the R-7 was truly operational. There was a pervasive sentiment in Washington that a totalitarian state with communal toilets could not pull off something so technologically complex, as Senator Ellender had stated. A similarly derisive disbelief had greeted the initial news in 1949 that Moscow had detonated an atomic bomb. "Do you know when Russia will build the bomb? Never," Truman had scoffed. When presented with incontrovertible evidence that the Soviets had indeed split the atom, Truman responded, "German scientists in Russia did it—probably something like that." Even after the USSR had further narrowed the atomic gap with a thermonuclear hydrogen bomb in 1953, Moscow was still the butt of American jokes. Russia couldn't possibly smuggle a suitcase bomb into the United States, went one popular punch line, because the Soviets hadn't yet perfected the suitcase.

Such perceived slights drove Khrushchev to push his scientists

Soviet leader Nikita Khrushchev (front row, center) and members of the Presidium in 1955. At the far left of the front row is Lazar Kaganovich, deputy premier, and next to him is Nikolai Bulganin, chairman of the Council of Ministers. To the right of Khrushchev are Soviet premier Georgi Malenkov and defense minister Klimenti Voroshilov. (© Bettmann/CORBIS)

Khrushchev often asked his son, Sergei, an engineering student, to accompany him on visits to the Soviet Union's rocket research facilities. (Photograph courtesy Sergei Khrushchev)

The tenacious Chief Designer of the Soviet missile program, Sergei Korolev, emerged from the Stalinist gulag to lead the Soviet Union into space. His identity was considered a state secret and his name never appeared in any news reports about missiles or satellites. (From the collection of Peter A. Gorin)

ABOVE LEFT: Major General John Bruce Medaris (left) was in charge of the U.S. Army Ballistic Missile Agency, which employed the top engineers from Nazi Germany's rocket program but which got scant support from the Pentagon. He is shown here with Brigadier General Holger N. "Ludy" Toftoy, who had whisked the scientists out of Germany in the summer of 1945. (*NASA Marshall Space Flight Center*)

ABOVE RIGHT: Charles E. Wilson, Dwight D. Eisenhower's secretary of defense, was the former president of General Motors and his charge was to bring down costs at the Pentagon. At his confirmation hearings, he said he did not see any conflict of interest in holding on to his GM stock because "what was good for the country was good for General Motors and vice versa." (*Department of Defense*)

Senator Stuart Symington of Missouri (left) chaired the Air Power hearings in April 1956 to investigate whether the United States was falling behind the Soviet Union. Wealthy, handsome, and quick to score political points, he was considered an early contender for the 1960 Democratic presidential nomination. Here he examines early American missile models with Lieutenant General James M. Gavin. (© *Bettmann/CORBIS*)

Korolev's archrival, the brilliant propulsion specialist Valentin Glushko, designed virtually all missile motors in the USSR. In 1938 he had provided damaging testimony against Korolev that was used as evidence at the trial that sent the Chief Designer to Siberia, but now they had to work side-by-side again. *(From the collection of Peter A. Gorin)*

A group photograph of many of the key participants in the Soviet missile and satellite programs. Korolev is in the front row, fourth from right. Others in the front row include Mstislav Keldysh (fifth from left), Leonid Voskresensky (seventh from left), R-7 Commission chairman Vasily Ryabikov (center, with legs crossed), deputy defense minister Mitrofan Nedelin (fifth from right, in uniform), deputy armaments minister Konstantin Rudnev (third from right), Valentin Glushko (second from right), and Vladimir Barmin (far right). *(From the collection of Peter A. Gorin)*

President Dwight Eisenhower and Vice President Richard Nixon were publicly united during the 1956 campaign. Behind the scenes, however, theirs was a strained relationship. *(AP/Wide World Photos)*

Walt Disney (far left) visits Wernher von Braun at the Redstone Arsenal before hiring him as a scientific adviser and host for the Tomorrowland segments of his new *Disneyland* television program. In these broadcasts, many Americans learned about satellite technology for the first time. *(NASA Marshall Space Flight Center)*

The powerful and staunchly anti-Communist Dulles brothers. Allen Dulles (left), the director of Central Intelligence, and John Foster Dulles, the secretary of state, set the tone for the Eisenhower administration's aggressive containment policies toward Moscow. *(© Bettmann/CORBIS)*

Richard Bissell was the man behind the CIA's top-secret U-2 and satellite reconnaissance programs. *(Central Intelligence Agency)*

The U-2 was used in reconnaissance missions over the Soviet Union, taking aerial photographs from as high as 70,000 feet. In the mid-1950s, it flew beyond the range of Soviet fighters or missiles, infuriating Khrushchev and helping to spur the development of missiles and satellites. *(U.S. Air Force Photo)*

Known as the "quiet man," Mikhail Tikhonravov was the introverted visionary behind the Soviet Union's satellite breakthroughs. *(From the collection of Peter A. Gorin)*

The world's first intercontinental ballistic missile, the R-7, seen here in its Tulip launch stand, was Khrushchev's bold gamble to redefine the arms race on his own terms. *(From the collection of Peter A. Gorin)*

The Soviet engineers called their first satellite PS-1, for *prostreishy sputnik* or "simple satellite." After its successful launch into orbit on October 4, 1957, it would be known simply as Sputnik. *(From the collection of Peter A. Gorin)*

President Eisenhower, encountering a hostile press corps, tried to downplay the military significance of Sputnik at an October 9, 1957, press conference. *(© Bettmann/CORBIS)*

Senator Lyndon Johnson of Texas (second from right), the Democratic majority leader, seized on the Sputnik scare to further his own ambitions, outmaneuvering Symington to hold hearings on "preparedness." Here he poses with a giant globe as Secretary of Defense Neil McElroy (in bow tie) and Deputy Secretary Donald Quarles point out the location of Sputnik's launch. Also looking on at far left is Senator Richard Russell of Georgia. *(AP/Wide World Photos)*

Eisenhower, in a morale-boosting address, tries to reassure Americans that the United States has not fallen far behind the Soviets. Here he displays a recovered nose cone from an American rocket shot into space. *(White House National Park Service Collection, courtesy Dwight D. Eisenhower Library)*

The navy's Vanguard missile was given priority for America's first post-Sputnik launch on December 6, 1957. However, the vehicle blew up on the launch pad, on live television, humiliating the United States in the eyes of the world. Newspapers called it "Flopnik" and "Kaputnik." *(NASA Headquarters—Greatest Images of NASA)*

America's second launch attempt would be a closely guarded affair. In the control room at Cape Canaveral in January 1958, Wernher von Braun (second from right, below) conferred with his colleagues as they waited for the weather to clear. *(NASA Marshall Space Flight Center)*

Finally, on January 31, 1958, America's first satellite, Explorer, is catapulted into space atop von Braun's Jupiter-C rocket. *(NASA Marshall Space Flight Center)*

harder to prove the critics in Washington wrong. But Korolev did not want to dwell on the time-consuming setbacks of the reentry problem. In the months required to completely redesign the warhead shield, momentum would be lost. Meanwhile, he had a rocket on its way to Tyura-Tam, and a record waiting to be broken. The Americans were surely not sitting idly by as the clock wound down on the International Geophysical Year. Korolev decided to turn the screws up a notch on his cautious fellow commission members. "I propose," he said airily, "that we put the question of the national priority of launching the world's first satellite to the Presidium. Let them settle the matter."

There was, of course, no "them" in the Presidium any longer. By September 1957 the Communist Party's ruling body was Khrushchev's personal rubber stamp. The dissenting voices were all gone, replaced by loyalists like the fainthearted but trustworthy Brezhnev, and newcomers like Andrei Gromyko, who replaced the turncoat Shepilov as foreign minister and was scheduled to travel to Washington in the first week of October to meet John Foster Dulles for the first time. The mutineers had been dealt with—though not in the customary Stalinist fashion that they had so ardently supported. In a testament to Khrushchev's reform of the dictatorship of the proletariat, not a single conspirator was shot or even arrested. Molotov was dispatched to Outer Mongolia, to serve out his sentence as the Soviet ambassador in dusty Ulan Bator. Kaganovich was appointed director of a remote potassium mine in the Perm province of the Ural Mountains. Malenkov was sent to manage the Ust-Kamenogorsk electric power station on the equally desolate Irtysh River in Kazakhstan, while Shepilov was dispatched to Kyrgyzstan to teach central Asian children the tenets of Marxist-Leninism. They were effectively banished into internal exile, but they would live out their natural lives.

Korolev's gambit had its desired effect. Opposition to the substitute satellite melted away almost as quickly as it had welled up at the previous meeting. The military men sat silent, and Glushko lowered his eyes in defeat. He may not have shared his rival's dreams of space conquest, but he certainly did not want to be the nail that stuck out. "Nobody wanted to be accused of dragging their feet," General Gerchik recalled, in the event that the United States did launch first and Khrushchev later came looking for answers and scapegoats. One after another, the commission members meekly raised their hands. The final decision was unanimous.

The only outstanding question, the launch date, was settled at the next meeting. On September 23, the same day as the Little Rock riot that spurred Eisenhower to action, the commission formally informed the Kremlin that PS-1 was scheduled for liftoff on October 6, 1957. It was official. The "Iron King," as the petrified staff at OKB-1 sometimes called Korolev, had won. The stubborn Chief Designer had finally gotten his shot at space.

• • •

Liftoff was scheduled for 10:20 PM on Sunday, the sixth of October, under the cover of darkness because American spy planes roamed the skies during the day. It also turned out that the late hour was ideal for PS-1 to attain its desired orbit. The launch itself would be strictly secret in case it failed, and Korolev took every precaution to ensure that Washington did not get wind of his intentions. In the huge assembly hangar not too far from the spartan little house the Chief Designer kept at Tyura-Tam, the R-7 lay prone on a train-sized dolly, its copper-clad exhaust nozzles burnished to a bright orange under its flared white skirt. It was model number 8k71PS, sixteen feet shorter than its predecessors, and technicians in surgical smocks were tinkering with the final modifications to its smaller, stubbier nose cone. The alterations gave the now ninety-six-foot rocket a stouter, more matronly look, but PS-1, Tikhonravov's tiny *prostreishy sputnik*, or "simplest satellite," did not require the same large and elongated thermal shield as a five-ton thermonuclear warhead that would reenter the searing atmosphere. The warhead's cumbersome radio-guidance targeting system had also been removed, shaving another four feet off the final package, since it too was no longer necessary. The satellite, after all, was not being aimed at an American city; with luck, it would never touch solid ground again. To achieve orbital velocity, Glushko's central sustainer core engine was being recalibrated to fire until it ran out of fuel, rather than to cut off at a predetermined point along a ballistic trajectory, and a new, more potent mix of hydrogen peroxide was being introduced to drive its turbo pumps faster.

Throughout the modifications, Korolev anxiously paced the enormous hangar like an expectant father in the delivery room. "Silence fell whenever the Chief Designer appeared," Colonel Mikhail Rebrov remembered. "Korolev was more exacting and strict than ever." Every so often he checked on Tikhonravov's baby, which sat on a felt-covered cradle in a sealed-off "clean room," Tyura-Tam's equivalent of a maternity ward. "Coats, gloves, it's a must," Korolev insisted, as he inspected the shiny satellite. Swaddled in a black velvet diaper, the little orb had spring-loaded antennae that dangled over the sides like electronic umbilical cords. To ensure its chances of survival, Tikhonravov had pressurized the sphere with nitrogen, a neutral gas that prevented corrosion. He had also installed a miniaturized climate-control system; it would heat or cool PS-1's innards to maintain a constant temperature of sixty-eight degrees Fahrenheit, which would ensure that its transmitters operated properly regardless of the external environment. No one knew for certain how the radio equipment—how anything man-made, for that matter—would react in the radiation-laden, zero-gravity vacuum of space, and that was another reason Korolev had insisted on the obsessive polishing of PS-1's thin aluminum skin. He did not want to risk the heat transfers or fluctuations that could result from an uneven surface, and wanted to make sure solar rays were reflected, not absorbed, by the gleaming shell. It was during one of these final, frenzied cleanings that the senior OKB-1 engineer Anatoly Abramov witnessed a typical Korolev moment. "I saw a crowd gathered around the satellite and I heard screaming," he recalled. "As I got closer I found myself at the receiving end of one of Korolev's famous tirades. I immediately realized what was wrong. The satellite stand was covered in felt to prevent scratching, but the felt had been tacked on with little nails rather than glued. The nail heads weren't actually protruding or touching the surface of the satellite, but it hadn't occurred to us that using tacks wasn't the brightest idea until Korolev rubbed all our noses in it."

The Chief Designer was next sent into a frenzy by a message that arrived from Moscow on the morning of October 2. Apparently an unscheduled meeting of the IGY was being convened in Washington on October 6. "There should be an American report of a satellite over the planet," the IGY's Soviet representative had cabled Moscow. In fact,

"Satellite over the Planet" was merely the title of the keynote speech. Either the Soviet representative got confused or something got lost in translation, and Korolev panicked.

"What does it mean?" he demanded, the color draining from his face. Were the Americans planning a launch? Were they planning to announce it on the sixth? "Maybe it's just a routine update," he tried to console himself. "Or maybe not," he said after a moment of anxious reflection, "maybe this will be a report of a fait accompli." Korolev was visibly shaken. A U.S. satellite might already be circling the earth by the time he attempted to launch PS-1. The idea of finishing second sent the Chief Designer into a state of profound agitation. He thrashed around his office, mumbling to himself, all the while clutching the worrisome communiqué. Get me the KGB, he finally roared.

Was it true? he asked, when the call was patched through. Were the Americans really about to launch a satellite? The duty officer at KGB headquarters did not know. A series of coded messages was exchanged between Moscow and the resident spies at the Soviet embassy in Washington. No, came the final answer. There were no early indications that the United States was planning any sort of launch.

Korolev, though, was far from relieved. What if the spooks were wrong? It wouldn't be the first time Soviet intelligence had missed signals. Korolev couldn't chance it. I'm moving up the launch date to October 4, he informed Vassily Ryabikov, chair of the R-7 State Commission.

This time he didn't bother to wait for an answer from Moscow.

• • •

The rocket was rolled out of its hangar the following morning. An overhead crane lifted the twenty-seven-ton empty shell—light and eminently more manageable without its warhead and full complement of fuel—and gingerly deposited it on a giant green erector-transporter that waited on rail tracks at the hangar door. Korolev, apparently still feeling the emotional pinch of the previous day's panic, patted his missile sentimentally. "Well," he told the assembled dignitaries, "shall we see off our first-born?"

A solemn procession began along the sandy mile-and-a-half-long berm that connected the assembly hangar to the launchpad, a tradition that would be repeated for every subsequent space launch and continues to this day. Heads bowed in silence, hands clasped behind their backs, the scientists, soldiers, and technocrats followed the locomotive that slowly, painstakingly pushed the R-7 on its transporter to the fire pit. A grainy and undated Soviet video captured the scene. In the front row, Korolev in a black leather jacket walked next to Voskresenskiy, his trusted chief of flight testing, looking like a portly French painter in the black beret that he used to seal liquid oxygen leaks with frozen urine. Farther back, the bemedaled generals, their olive green uniforms matching the military paint job on the 150-foot-long transporter. Behind them, Glushko, Ryabikov, and Rudnev, the deputy minister for military-industrial works. Then, bringing up the rear, the rest of the bureaucrats and lesser designers. In the video, no one is talking, and faces seem grim. The camera pans away to reveal a tableau of windswept dunes and a pair of camels on a ridge—though these have almost certainly been spliced in for exotic effect since it was highly unlikely that Kazakh herders were permitted to wander freely around Tyura-Tam.

Fifty minutes elapsed before the R-7 reached the launchpad, and the huge hydraulic boom on the transporter began to inch upward. Slowly, over the next hour and ten minutes, the rocket was raised into the waiting arms of the Tulip launch stand. When at last it had been fully righted, the transporter boom lowered it and the Tulip's petals closed around its waist like a vice. The R-7 was now suspended in midair, its thrusters hanging just below ground level over the 120-foot-deep, five-football-fields-wide concrete apron of the fire pit. But before fueling could begin, it still had to be tested one last time. It was a shortcoming of horizontal assembly, a time-consuming extra step that the Americans had skipped by building their towering new hangar at Cape Canaveral several dozen stories high so that U.S. missiles could roll out already tested and in the vertical position.

Marshal Nedelin, in particular, was unhappy with the Soviet arrangement. He was going to head the Strategic Rocket Forces, and in the event of a nuclear attack, precious time would be lost running unnecessary diagnostics. An ICBM's retaliatory value depended largely on how quickly it could be fired, and the R-7 was proving painfully slow to get off the ground. Nor could problems be fixed once the missile was fueled, due to the risk of explosion.

Nedelin paced impatiently throughout the morning, glancing disapprovingly at his watch as the technicians checked connections and valves and electrical circuits. Sometime during the diagnostic tests—there are conflicting accounts as to precisely when—a malfunction with the satellite was uncovered. One of its silver zinc batteries was leaking electrolytes, and there was a disruption in the current. "Technical banditry," howled Rudnev, the man who had assured Chertok that no one would be sent to Siberia if the R-7 failed. But now, in the heat of the moment, he wanted heads to roll for the perceived sabotage. Korolev, however, was uncharacteristically calm. "Let's not make a fuss," he consoled the highly agitated deputy minister. "There is still time to make the necessary corrections."

It was not until shortly before six the following morning, on Friday, October 4, that fueling could begin. By then, many of the launch crew had fallen ill from spending so much time in the unseasonably cold weather. An Arctic blast had descended over the Kazakh steppe from Siberia, bringing howling winds and freezing temperatures, but the personnel at Tyura-Tam were still dressed for the broiling summer. Huddled around a makeshift shack that served moldy salami and stale pastries but no hot tea, the soldiers and technicians shivered and cursed. "OK, dear," said one, addressing the missile. "Fly away and carry our baby into space. Or at least crash. Just fly away, and don't stay here," he added, dreading the prospect of the additional days it would take to drain and dismantle a stalled rocket.

Rail tankers containing 253 tons of kerosene and supercold liquid oxygen pulled up to the hinged girders of the Tulip, and soldiers heaved huge hoses onto cables and pulleys that hoisted them up to the R-7's intake valves. The troops manning the fueling operation wore no protective clothing other than gloves, and clouds of cryogenic condensate descended on them through the bleed valves that hissed frozen oxygen vapors as they pumped a small amount of liquid oxygen to cool and pressurize the rocket's plumbing.

The nearly minus-300-degree liquid oxygen evaporated at an alarmingly rapid rate, which was why the R-7 had to be filled shortly before takeoff and its tanks constantly topped off, and could not be stored ready for firing like future generations of ICBMs that would use storable propellants. The combustible mist infused the soldiers' hair and clothes; eventually, after several horrific cases of people igniting, the Soviets would adopt more stringent safety precautions. But during the early R-7 launches, caution was not a concern.

The fueling process lasted five excruciating hours, the soldiers carefully distributing the propellant into each of the missile's ten integral tanks to maintain weight equilibrium. Compressed gases like nitrogen and liquefied hydrogen peroxide were then pumped under high pressure into the turbos that would drive the fuel pumps. Throughout the arduous process, Nedelin once again must have watched the clock with alarm and dismay. The next war would be an instantaneous conflagration, won or lost not in a matter of days or months but hours and minutes. In such a conflict, when missiles could cross continents and oceans in the time it took to load a bomber, five hours was an eternity. The Americans were already talking about designing a new storable solid-fuel rocket that could be ready to launch in less than five minutes, and here Nedelin had to wait a day and a half just to top off the tanks. The very same soldiers fueling the R-7 would have to fire it in the event of a war, and unless they picked up the pace, the R-7 risked being taken out while it was still on the ground.

Korolev, however, ignored the impatient rumblings of the military observers. "Nobody will rush us," he instructed his engineers. He had come too far to make a mistake now. He had waited twenty years for this moment, sacrificed his marriage to Ksenia, his health, even his freedom during the purges to work on rockets. He could wait a few more hours. "We will launch at 22 hours and 28 minutes," he announced.

· · ·

"T minus ten minutes," blared the loudspeaker, as Korolev, Voskresenskiy, and the other R-7 State Commission members filed into the underground control bunker 200 yards from the launchpad. Above them, powerful spotlights illuminated the frost-covered rocket, which glistened in the night like a giant icicle. Steam hissed from its bleed valves, enveloping the launch stand in thick, billowy clouds bisected by sharp beams of light.

At 10:20 PM, the rocket's automated guidance systems were switched on, and its inertial gyroscopes began spinning, emitting a low hum. Inside the crowded bunker, the military operators manning the dimly illuminated panels and dials of the various control stations scanned their indicators for signs of trouble. Almost immediately, a warning signal on the Auxiliary Systems panel started flashing. It was the fuel tank sensor in one of the peripheral boosters. The level of liquid oxygen was low. All eyes turned to Korolev. Was it serious? Should they abort? Korolev and Voskresenskiy exchanged meaningful looks and huddled in a whispered conference with the two ranking military launch commanders. It wasn't critical, Korolev decided. They would proceed with the countdown.

Voskresenskiy returned to the helm of one of the bunker's two periscopes and stared out through the viewer. The R-7 seemed fine. He flashed the Chief Designer a brief, helpless smile. He and Korolev had just made their final decision. The launch was now out of the scientists' hands, an entirely military operation, and as civilians they were henceforth just spectators.

"One minute to go," announced Colonel Aleksandr Nosov, swiveling the second periscope like a submarine commander. This was now Nosov's show, and though he was aiming at space, the launch would be treated like a regular ballistic missile training exercise. "Key to launch," he ordered, and Lieutenant Boris Chekunov, the "button man," inserted the key that controlled the circuit breaker on the firing switch. "Key on," Chekunov responded.

"Roll tape." The telemetry readouts began rolling off the printer like a stock market ticker tape. "Purge the system," Nosov called out ten seconds later. Inside the rocket, compressed nitrogen was blasted through the engine feed lines to flush out any gaseous residue from the fueling and testing. "Key to drainage." Chekunov flipped the switch, and all the bleed valves closed. The hissing and steaming abruptly ceased, and the vapor clouds around the rocket disappeared as the last of the feed lines that topped off the evaporating liquid oxygen was automatically disconnected. Two minutes passed before Nosov issued his next command: "*Pusk,*" or "Launch."

Chekunov pressed the launch button, starting the automated sequence. Inside the R-7, compressed nitrogen rushed into the propellant tanks, pressurizing them to the bursting point. The umbilical mast with the ground electrical connections retracted and the missile switched to onboard battery power.

"Roll tape two," Nosov commanded ninety seconds later. Every ground receiving station in the Soviet Union was activated to full power, ready to track the rocket. It was now 10:28 PM. Inside the R-7, valves opened, and the turbo pumps began sucking thousands of gallons out of the propellant tanks. "Ignition," called Chekunov, reading the flashing light on the panel in front of him. From their periscopes Voskresenskiy and Nosov could see a cloud of orange smoke envelop the rocket, as flames poured out of the thirty thrusters. But the fire was languid and lazy, dancing, directionless. "Initial stage," Nosov called out. The engines were only warming up; the turbo pumps that fed fuel to the combustion chambers were operating at a fraction of their capacity. This was normal and followed after a few seconds by a ground-shaking roar. "Primary stage," Nosov shouted, as the R-7 went to full thrust. An ear-splitting din, like the sound of lightning as it strikes, penetrated the bunker's thick concrete walls, and the light coming through the periscopes' viewfinders was blinding as the flames shooting out of the rocket intensified to white-hot jets of superheated gas. They slammed into the bottom of the fire pit with such force that updrafts propelled them back up the sides of the missile 120 feet above. For a split second, the rocket sat there burning itself alive, and then it slowly rose from the pyre. "Liftoff, liftoff," Nosov screamed, as a million pounds of downward pressure pushed the Tulip's hinged pedals open and the R-7 was released.

In the eight seconds it took the 280-ton missile to climb the first 1,000 feet, an alarm indicator had silenced the cheers in the control room. The engine of one of the peripheral blocs, the same side booster that had registered low liquid oxygen levels, had been late achieving full power. The rocket had still taken off normally, but that didn't mean it wasn't a sign of trouble to come; the problematic booster might still suffer a critical failure before it separated. The seconds were ticking by quickly, though, and there was nothing anyone could do now but monitor the display panels and stopwatches and hope for the best. At sixteen seconds, another alarm indicator began winking. The Tank Depletion System, which ensured that propellant flowed evenly to all the combustion chambers, had malfunctioned. The engines weren't burning fuel uniformly, which could affect the rocket's course and, more important, its speed and preprogrammed cutoff time. Now everyone was seriously worried. The glitches were piling up fast, and no one had forgotten the disaster that had occurred at the ninety-eighth second of the first R-7 flight.

At 116 seconds a fiery cross appeared thirty miles above the Tyura-Tam test range. The four side boosters had jettisoned, creating the biblical effect, and miraculously the separation had occurred exactly on schedule. Relief swept through the control room. Only the central sustainer core was now firing, which meant that fewer things could go wrong. Glushko's reconfigured engine had enough fuel for two more minutes of flight. Then they would know.

The control bunker was subdued; there were too many generals and colonels and deputy ministers present for the young lieutenants in the launch crew to display their emotions. But in the assembly hangar, where most of the civilian scientists and engineers listened to the action on a loudspeaker, it was a different story. There, emotions ran high; whoops and cheers greeted milestones, while announcements of glitches were met with moans and groans.

For the next two minutes, all eyes were riveted on the clock. Then the loudspeaker sounded. "Main engine shut down." A distressed murmur reverberated through the hall. The engines had run out of fuel at 295.4 seconds. That was more than a full second early, a result of the Tank Depletion System malfunction. Slide rules were whipped out and calculations hastily performed. Would the early cutoff affect escape velocity? The R-7 was supposed to be traveling at just over 8,000 meters per second—roughly 18,000 miles an hour—but it was making only 7,780 meters per second. It was also five miles lower than it should be, at 142 miles in altitude instead of 147 miles. Would it be enough to orbit? Another 19.9 seconds passed before the next announcement. Meanwhile, momentum had carried the missile, still traveling at twenty-three times the speed of sound, another one hundred miles higher. "Separation Achieved." Inside the R-7's nose cone, pneumatic pistons rammed PS-1's steel cradle, pushing it away from the spent booster. A spring-loaded mechanism popped off PS-1's conical cover, and the sphere hurtled into the blackness of space.

At 325.44 seconds into the flight, Nosov issued his last command. "Open the reflectors." A plate on the central booster jettisoned, exposing prismlike mirrors on the rocket's casing. Korolev had installed the reflective material, knowing that the ninety-foot central stage would follow PS-1's celestial path like the blazing trail of a meteor, and he wanted to ensure that it too would be visible from earth as it circled the planet just behind the satellite.

But was PS-1 really in orbit? Had the little orb survived the violent shaking and vibrations of takeoff? Had it overheated during its ascent, succumbing to the friction of slamming through the dense lower atmosphere at nearly 25,000 feet per second? Had the thin cover shields held? Everyone rushed to the communications van parked outside to find out. The van sprouted an array of antennae tuned to the two frequencies of PS-1's twin transmitters. Inside the van, both operators hunched over their dials, cupping their headphones. "Quiet," one of them yelled. "Be quiet." So many people were pressing against the vehicle, clamoring for information, that the two operators couldn't hear anything. Then, one of them raised an exultant arm. "We have the signal," he shouted. "We have it."

Celebration erupted: dancing, laughing, hugging. Grown men cried and kissed one another. Glushko and Korolev embraced, their clashes momentarily forgotten. "This is music no one has ever heard before," the Chief Designer cheered. Even the rigid military engineers inside the control bunker rose out of their seats in a rare display of emotion, though Chekunov, the young lieutenant who had pressed the launch button, would later recall that none of them would truly understand what had just happened until much later.

Reports now started trickling in from the Far Eastern tracking stations. One after another, they were acquiring PS-1's signal. It was on course, and its orbit seemed to be holding steady. Only a relatively minor altitude loss of fifty miles was reported. Once more cheering and shouting erupted, because that meant that the early engine cutoff had not had disastrous consequences after all. Already some of the State Commission members were reaching for the phones, ready to call Moscow with the good news. Korolev, though, was surprisingly subdued and silent. "Hold off on the celebrations," he finally counseled. "It could still be a mistake. Let's wait to hear if we can pick up the signal after a complete orbit."

For an hour and a half they waited, smoking, pacing, and fidgeting. When the appointed time for PS-1 to reappear over Soviet territory came and went in silence, a deathly stillness descended on the anxious crowd assembled in the huge hangar. A sense of foreboding suddenly gripped the scientists. Maybe PS-1 had continued to lose altitude and had burned up in the atmosphere. Maybe they had failed after all.

At a few minutes after midnight, one of the westernmost tracking stations in the Crimea picked up something. At first faintly and with static, and then louder and clearer: BEEP, BEEP, BEEP.

Amid the pandemonium, Korolev turned to his fellow State Commission members. Now, he said triumphantly, we can call Khrushchev.

8

BY THE LIGHT OF A RED MOON

General Bruce Medaris greeted October 4, 1957, with the giddy anticipation of someone expecting a new lease on life. The day, he felt sure, would mark a turning point for his besieged Army Ballistic Missile Agency—perhaps even offer a reprieve for his own troubled military career.

The source of this uncharacteristic optimism was the scheduled arrival that Friday morning of yet another high-ranking delegation from Washington. This time, though, Defense Secretary Charles Wilson would not be among the visiting brass, quibbling about the guest cottages. Wilson's reign of terror was over.

Engine Charlie—the man who had sidelined ABMA and tried to put it out of the missile and satellite business, a man so hated in Huntsville that some rocket scientists had once burned his effigy in Courthouse Square—was quitting. Whether Colonel Nickerson's whistle-blowing scandal and allegations of corporate cronyism had influenced his decision to return to Detroit to devote himself to automotive and charity work, Medaris did not know. Nor did he care. All that mattered was that his nemesis would be out of office by October 8. "We could not shed a single tear over Mr. Wilson's departure," Medaris later reminisced. "It was our strong feeling that his tenure had been characterized, to put it charitably, by a complete lack of imagination."

All of Huntsville, apparently, was of the same mind. The town had more than doubled in size since von Braun and his German engineers had moved into their brick ramblers in new suburbs with nicknames like "Sour Kraut Hill," and Huntsville's fate was now inextricably linked to ABMA and its high-tech marvels. The jobs of five thousand skilled workers and much of the local economy had hung in limbo since Wilson's November 1956 edict had effectively robbed the army of the big missile brief, and the uncertainty had devastated morale and depressed the once-booming real estate market. Huntsville, which had dubbed itself "Rocket City, USA," was learning the harsh reality of the military-industrial complex: with the stroke of a pen in Washington, entire communities could be wiped out as quickly as they were created.

To keep his company town afloat and his rocket team intact, Medaris had waged increasingly inventive bureaucratic guerrilla campaigns that were beginning to take their toll on his standing with the power brokers at the Pentagon. The embarrassing disclosures of alleged favoritism at the Nickerson court-martial had won the Jupiter intermediate-range ballistic missile program a temporary stay of execution; ABMA could continue doing limited research on the missile while the Pentagon decided whether to cancel the project entirely. Unfortunately, ABMA had few friends at the defense secretary's office, and the army IRBM was still on Charlie Wilson's chopping block.

Medaris's tireless lobbying to land a role for the army in satellites was also becoming an irritant. He had loudly and repeatedly questioned the selection of an inexperienced civilian team to launch the navy's Vanguard satellite, the official U.S. entry in the IGY competition, and hinted darkly at conspiracies in high places. He had also pushed his boss, James Gavin, a hard-nosed former paratrooper in charge of Army Research and Development, to lodge formal but futile appeals with Quarles and Wilson on ABMA's behalf.

Like Korolev, Medaris had simply refused to take no for an answer, and like the Chief Designer, he had not been above using a little subterfuge. The similarities were not that surprising, given that the two men had been raised in almost identical circumstances by strong-willed, single women who had challenged the chauvinism of their times. Medaris's mother had also divorced young and left her son with her parents while she pursued a career, eventually becoming the comptroller of a midsize manufacturing company and one of the most senior female executives in Ohio. It was in his grandmother's home that Medaris first displayed his resistance to authority, "timing his comings and goings so that Grandmother LeSourd didn't ask hard questions." From his industrious mother, Jessie, he learned the value of entrepreneurship, taking a part-time job at the age of eleven sorting mail at the local railway station. By twelve, he was driving a cab on weekends (driver's licenses were not yet required in Ohio), and at fourteen he was working full-time as a uniformed conductor on the Springfield Street Railway System on the 3:30-to-midnight shift. By the time the stock market collapsed in 1929, Medaris had accumulated over one hundred thousand dollars in his trading account. Left with sixty-nine dollars after the crash, he bought himself a new suit and started all over again.

Medaris was no stranger to adversity, and he was not a quitter. Regardless of what the Pentagon said, he would not abandon his satellite quest. And so, with Gavin's tacit compliance, he had "bootlegged" the Jupiter C. Ostensibly an experimental vehicle to develop a new form of ablative nose cone whose heat shield peeled off in layers, the C in reality was a souped-up Redstone whose added upper stages were suspiciously similar to the army's rejected satellite booster design. "We must make it perfectly clear," Medaris instructed von Braun and his staff, "that we did not carry forward a program which we had been denied, that the work was carried on because that was the best way to make a reentry missile and the two happened to fit together." Justified by the need to simulate the speed, friction, and trajectory of big nuclear-tipped missiles on atmospheric reentry, the Jupiter C's ulterior purpose was to keep ABMA unofficially in the satellite sweepstakes, since the navy was stumbling badly. Vanguard was so behind schedule and over budget that Eisenhower had considered canceling it, while the Jupiter C had set U.S. altitude and distance records, soaring 662 miles high over a 3,335-mile arc. Even so, Medaris had been unable to gain any traction with Wilson in repeated pleas to at least consider designating the army as a backup for the navy satellite effort. "In various languages our fingers were slapped and we were told to mind our own business," he recalled. "Rightly or wrongly, we were convinced that during Wilson's regime the Army had consistently been pushed aside."

ABMA had gotten a raw deal under Engine Charlie. But now his replacement, Neil H. McElroy, was coming to Huntsville. Medaris hoped to get a fairer hearing from the new secretary-designate, who was by all accounts a fair and forward-thinking man. At fifty-three, he was almost exactly Medaris's age, and a fellow Ohioan to boot. Like Medaris, he had a reputation for speaking his mind, and he had a midwesterner's impatience with Washington's insular ways. McElroy, in fact, had accepted Eisenhower's invitation to join his cabinet only on the condition that he serve no more than two years. Any longer, he argued, and he would risk succumbing to the temptations of political power.

Medaris felt certain he could reason with such a man, a son of small-town schoolteachers, a full-scholarship Harvard graduate. "Our whole organization was thoroughly fired up," he recalled. "We hoped that with a fresh and uncommitted mind, [McElroy] would grasp the significance of our story. We were determined to give him our frank feelings, backed by facts and figures, as to our record for delivering what we promised, when we promised, and for the money originally stated."

If Medaris had one reservation about McElroy, it was that he was yet another moneyed representative of big business, the president of the household goods giant Procter & Gamble. But at least he wasn't from the incestuous defense establishment, intent on feathering his company's nest with government contracts. In any event, Medaris would have the incoming secretary in Huntsville for a full twenty-four hours to bend his ear and make his case before the Washington hyenas got to him.

By the time McElroy's plane touched down at the Redstone Arsenal airstrip at noon on October 4, General Gavin had already been working on the secretary-designate during the flight, priming him for Medaris's pitch. The hard sell, though, was to take place that evening at the Officers' Club, over dinner and drinks and an outpouring of southern hospitality at a reception in McElroy's honor.

Wilbur Brucker, the secretary of the army, and General Lyman L. Lemnitzer, the army chief of staff, were in attendance, as were Huntsville's eager-to-please town fathers. Will Halsey, one of the community leaders, remembered the room being "so heavy with top brass that it seemed like two-star generals were serving drinks to three-star generals."

As the cocktails were being poured and the secretary's favor curried,

ABMA's public relations officer, Gordon Harris, abruptly burst into the bar. Clearly agitated, the young officer rudely interrupted McElroy and grabbed Medaris. "General," he stammered, too loudly for discretion, "it has been announced over the radio that the Russians have put up a successful satellite!"

For a moment, the room was deathly quiet, so that only the soft sound of background music could be heard. "It's broadcasting signals on a common frequency," Harris went on, as hushed murmurs began rippling through the gathering. "At least one of our local 'hams' [amateur radio operators] has been listening to it."

Then dozens of voices erupted in a spontaneous outburst of anger and pent-up frustration. "General Gavin was visibly shaken, and understandably so," an aide to Secretary Brucker later recalled. Gavin, only days earlier, had tried to persuade Wilson one last time to take the Jupiter C as a backup for the problem-plagued Vanguard. Now he cursed Engine Charlie's lack of foresight.

"Damn bastards" was all Medaris said, and it was unclear whether he was referring to the Soviets or his own government overseers. Whichever the case, he was stunned. How could the Russians have done it? It was impossible. Only the week before, he had laughed when Ernst Stuhlinger, one of von Braun's top engineers, had pleaded for him to approach Quarles because he was "convinced" that the Soviets were planning a launch. "Now look," Medaris had replied, "don't get tense. You know how complicated it is to build and launch a satellite. Those people will never be able to do it. Go back to your laboratory and relax."

Medaris had always presumed that he was in a race with the air force and the navy, not with the USSR. Like most Americans, he thought the Russians were boors: primitive, simple, crude. How could they pull off something like this? Medaris, for once, was speechless. Yet in underestimating the Soviet Union's technical potential, he had made the same mistake as Senator Ellender and all the others who had laughed at Moscow's crummy cars and shoes. What Medaris, like most Americans, failed to understand was that conditions that made communism wholly unsuited as a producer of quality consumer goods made it an ideal system for promoting major scientific breakthroughs. The state could never compete with private businesses making sneakers, tennis racquets, or transistor radios. But no corporation could muster

the vast resources, strict discipline, and unlimited patience that were required of huge scientific undertakings like the Manhattan Project, or
the creation of a satellite-bearing ICBM. Stuhlinger and von Braun, as
veterans of the state-run V-2 program, understood this and knew that
science thrived under totalitarian regimes, even if free speech and commerce did not.

Medaris had never grasped the dichotomy, and now, as the shock settled in, he didn't look the least bit relaxed. But it was the usually unflappable von Braun who appeared most emotional. "Von Braun started to
talk as if he had suddenly been vaccinated by a Victrola needle,"
Medaris later recalled. "In his driving urgency to unburden his feelings,
the words tumbled over one another."

"We knew they would do it!" von Braun exclaimed, his Teutonic
Texas twang rising to a fevered pitch. "We could have done it two years
ago," he cursed, launching into the story of how Wilson had been so
suspicious that the army might "accidentally" launch a satellite ahead of
the navy, igniting an interservice war, that he had ordered Medaris to
personally inspect the Jupiter C booster to ensure that the top stage was
a dud. (The precaution, as it turned out, had been unnecessary. "There
was no chance of an unauthorized attempt," Stuhlinger later recalled.
"We had our orders, and von Braun was very strict about following
orders.")

Office politics had denied von Braun his lifelong dream. Unlike Korolev, he had been obsessed with the conquest of space since early childhood. He had sold his soul—first to the Werhmacht, then to the Nazis,
and finally to the U.S. Army—to pursue his quest. He had endured
Hitler, Himmler, five long years of purgatory in the hot Texas sun. All
so that he could pursue his dream. And now, because of some idiotic bureaucratic imperatives, someone else had beaten him to it. Von Braun
very nearly exploded with anger and frustration. "For God's sake cut us
loose and let us do something," he implored McElroy. "We have the
hardware on the shelf."

Medaris must have swallowed hard. In his overexcited state, von
Braun had let it slip that ABMA had quietly diverted two Jupiter C
rockets from the nose cone testing program and put them in cold storage in anticipation of the navy not delivering Vanguard. A lot of people
had been complicit in the scheme, but the unauthorized misplacement

of millions of dollars of Pentagon property was not something one necessarily wanted to spring on the secretary of defense before he was even sworn in. ("It was imprudent to admit we had retained those rockets," Medaris would later confess to congressional investigators.) McElroy, though, made no comment, perhaps because von Braun did not pause for breath. "Vanguard will fail," he went on, with a certainty that verged on arrogance. "We can put up a satellite in sixty days, Mr. McElroy. Just give us the green light and sixty days."

"Ninety days," Medaris quickly interjected. Two months was pushing it. Von Braun, though, kept repeating his original figure. "Just sixty days."

"No, Wernher." Medaris finally pulled rank. "Ninety days."

But McElroy was in no position to make any spot decisions. He still had to be confirmed by the Senate, which was controlled by the Democrats, who were likely to develop a sudden and intense interest in the subject of space.

Reports on the Soviet satellite now began trickling out on the radio and television at the Officers' Club. Harris, the harried PR officer, announced that ABMA's communications team had also captured its signal. "It beeped derisively over our heads," he said. Western news agencies in Moscow had by now hastily translated the official TASS press release, which included technical details of the orbiting craft. The Soviets were referring to it as *Iskustvenniy Sputnik Zemli*, or Artificial Satellite of the Earth. American broadcasts were simply calling it Sputnik, the generic Russian term for satellite. ABMA's scientists now clustered around Harris, bombarding him with questions. What were Sputnik's parameters? What was its orbit? How big was it? When word spread that it weighed 184 pounds, people shook their heads in disbelief. Must be a mistake, they said. Someone must have misplaced a decimal point. Vanguard's satellite payload was only 3.5 pounds because the navy's slim booster produced a mere 27,000 pounds of thrust. Even von Braun had never proposed anything larger than 17 pounds as the payload for his 78,000-pound-thrust Jupiter C. How could the Soviets put up a satellite ten times heavier? Plainly the media had got it wrong. But if the press reports were accurate, the military implications of a missile powerful enough to orbit such a weighty cargo were frightening. It would have to generate hundreds of thousands of pounds of thrust, possibly as much as half a million, some of the scientists ventured. (Not even their wildest guesses, however, approximated the R-7's 1.1 million pounds of lift.)

Throughout all the frenzied speculation, Medaris and von Braun kept hammering away at McElroy, who must have felt as if he were being baptized by fire. "Missile number 27 proved our capabilities," Medaris pressed, referring to the Jupiter C shot that had reached approximately the same altitude as Sputnik was currently circling overhead. "It would have gone into orbit without question if we had used a loaded fourth stage. The hardware is in hand, and so the amount of money needed to make the effort is very small," Medaris said, continuing his hard sell. "I believe we have a 99% probability of success."

Give us $12.7 million and the go-ahead, Medaris pleaded. "We felt like football players begging to be allowed to get off the bench and go into the game to restore some measure of the Free World's damaged pride," he recalled later.

Sputnik, as Medaris and von Braun had almost immediately grasped, was ABMA's ticket out of the doldrums, an opportunity to be seized. Surely the administration would have no choice but to respond to the Soviet challenge, and ABMA was the nation's best bet to even the score. "When you get back to Washington, and all hell breaks loose," von Braun told the secretary in one last sales pitch as McElroy was boarding his plane the next morning, "tell them we've got the hardware down here to put up a satellite any time."

Medaris, in fact, had already ordered von Braun to secretly start preparing for launch. He was so confident that the political fallout from Sputnik would spur the White House to action that he had skipped waiting for the green light. What the maverick general did not realize, however, was that his commander in chief would have decidedly different ideas.

• • •

The debacle in Little Rock had shaken Dwight Eisenhower. For the first time in his presidency, a majority of the American people—64 percent, according to a Gallup survey—had disapproved of the way he had handled a crisis. His trademark calm and restraint had abandoned him in the wake of Governor Faubus's impudence, and many voters felt he had overreacted by sending troops to Arkansas. Not surprisingly, the polls skewed most unfavorably in the South.

While the immediate crisis was over (though paratroopers remained posted outside Central High School), Faubus was apparently still weighing heavily on the president's mind when he returned from his three-week vacation in Newport in the waning days of September. "Dear Dick," he wrote Nixon on October 2, extending an olive branch to the marginalized vice president, whose calls for a more forceful stand on integration he had long ignored. "I had been hoping to play golf this afternoon. . . . If you already have a game, please don't think of changing your plans because mine are necessarily uncertain because of the stupidity and duplicity of one called Faubus."

Nixon, as it turned out, did not have a golf game planned for the middle of the workweek and jumped at the rare opportunity to join his usually distant boss for a 1:00 PM tee time at the Burning Tree Country Club. The two had hardly seen each other over the past several months, owing to Eisenhower's extended vacations and Sherman Adams's tight control over entry to the Oval Office. Only John Foster Dulles had unfettered access to the president. Nixon, like everyone else, had to go through the chief of staff. The restrictions grated. "Sherman Adams was cold, blunt, abrasive, at times even rude," Nixon vented in his memoir. The vice president was increasingly clashing with Adams, since he was trying to carve out a more meaningful role for himself in Eisenhower's second term, especially in the realm of foreign policy. He was now a likely front-runner for the 1960 Republican presidential nomination, and he needed to raise his profile. Maybe he could persuade Ike to send him on some state visit.

But the president didn't seem interested in talking shop. "Golf in Newport was enjoyable," he remarked amiably. "I got to the point where I was hitting the ball as long as I ever did." His putting, however, Eisenhower complained, had suffered "a corresponding slump."

Eisenhower's insouciance was partly an act. The president was tired and worn down by a summer of squabbling with Stuart Symington and the other Democrats in Congress over defense spending, and he was increasingly worried by some troubling numbers coming out of the Commerce Department. The economic boom he had inherited during his first term, when more than one million families a year were moving up into the middle class in what *Fortune* magazine called "an economy of abundance," appeared to be faltering. Unemployment figures for August were showing a sharp rise. The real estate and stock markets had cooled considerably. Consumer confidence indicators were down. And tax revenues were coming in at a disappointing $72 billion, $4 billion below projections. Prosperity and fiscal prudence were pillars of the administration's platform, and Eisenhower, at Charlie Wilson's suggestion, had ordered sweeping military cuts in July in an effort to trim half a billion dollars from the $3.5 billion monthly defense bill. Already the Democrats were howling that his policies favored the rich while putting the country at risk. "What the hell good is it to be the richest man in the graveyard?" Symington had snapped. And now Ike faced the agonizing possibility of a looming recession to further complicate his budget balancing act.

"The developments of this year," he wrote in a diary entry on September 13, 1957, a week before Little Rock, "have long since proved to me that I made one grave mistake in my calculations as what a second term would mean to me in the way of a continuous toll upon my strength, patience, and sense of humor. I had expected . . . to be free of the many preoccupations that were so time consuming and wearing in the first term. The opposite is the case. The demands that I 'do something' seem to grow."

At nearly sixty-seven years of age, with a heart attack and stomach surgery on his recent medical record, Ike simply didn't have the stamina he had once had. And so, on Friday, October 4, he decided to take a break and spend a recuperative four-day weekend at his beloved farm in Gettysburg, Pennsylvania.

With its prizewinning herd of Black Angus beef cattle and historic battlefields, the putting green he had installed just outside his patio doors, and the reassuring scent of his wife Mamie's rhubarb pies wafting out of the kitchen, Eisenhower cherished the farm above all his other possessions. He never took more than a skeletal staff to intrude on his privacy at Gettysburg, and that is perhaps one reason why there is no record from any White House aides as to how the president reacted when told that night that the Soviet Union had launched a satellite. What is known, according to the space historian Paul Dickson, is that the next morning the president of the United States played golf for the fifth time that week.

In Eisenhower's absence, it was John Foster Dulles who drafted the official White House response to Sputnik. The launch was "an event of considerable technical and scientific importance," Dulles allowed in an October 5 statement. "However, that importance should not be exaggerated. What has happened involves no basic discovery and the value of a satellite to mankind will for a long time be highly problematical. The Germans had made a major advance in the field and the results of their efforts were largely taken over by the Russians when they took the German assets, human and material."

The gist of the press release was clear. Sputnik, as far as the White House was concerned, was not a big deal. If anything, it was a feat of Nazi engineering, not Soviet know-how—never mind that the Germans in question were beavering away in Huntsville, not Moscow. The tone thus set, administration officials lined up to spin the news. Sputnik was "without military significance," said the White House aide Maxwell Rabb. "A neat technical trick," shrugged Charlie Wilson. "A silly bauble," scoffed Eisenhower's adviser Clarence Randall. Sputnik did not come as the least bit of a surprise, Press Secretary Jim Hagerty assured the world. America was not interested in getting caught up "in an outer space basketball game," Sherman Adams announced. The satellite was a useless "hunk of iron that almost anyone could launch," growled Admiral Rawson Bennett, Vanguard's commanding officer.

Loyal Republican lawmakers added their voices to the chorus of skepticism. Sputnik was nothing more than "a propaganda stunt," said Senator Alexander Wiley of Wisconsin. It was like a "canary that jumps on the eagle's back," declared Representative James Fulton of Pennsylvania, apparently insinuating that the Soviets were hitchhiking off American technology.

But much as the administration tried to downplay the significance of the Communist breakthrough, the media decided differently. Sputnik was a big story—a very big, shocking, scary story. "Listen now for the sound that will forever more separate the old from the new," intoned NBC, broadcasting Sputnik's beep on Saturday, October 5. "Soviet Fires Earth Satellite into Space," the *New York Times* trumpeted in the sort of six-column-wide headline usually reserved for declarations of war. "Sphere Tracked in Four Crossings over U.S."

From the journalistic perspective, Sputnik had everything going for it: a historic milestone of human evolution, the element of surprise, the sting of defeat, and frightening ramifications as CBS's Eric Sevareid somberly informed viewers in his October 6 telecast.

> Here in the capital responsible men think and talk of little but the metal spheroid that now looms larger in the eye of the mind than the planet it circles around. Men are divided in their feelings between those who rejoice and those who worry. In the first group are the scientists, mostly, in raptures that the nascent, god-like instinct of *Homo sapiens* has driven him from his primordial mud to break, at last, the bound of his earth. Those who are worrying tonight know that the spirit of man has many parts: and part of his spirit is not in space; it has not even reached the foothills. And so broken men still lie in Budapest hospitals because a form of ancient tyranny finds free thought a menace; and in mid-American cities bodies and hearts bear bruises because this part of the human spirit still fears and hates what is different, even in color. The wisest of men does not know tonight whether man in his radiance or man in his darkness will possess the spinning ball.

America had been bested on the international stage, and editors across the land now salivated at the prospect of finding someone to blame for the sluggishness and complacency of the U.S. satellite program. The administration's underwhelmed response smacked of sour grapes and made it an appealing target for the nationwide editorial witch hunt. Nothing, after all, sold newspapers like the old-fashioned whiff of incompetence and scandal.

Sputnik contained one final element that no ambitious newsman could resist: fear. The missile that had lofted Sputnik into space had also shattered America's sense of invulnerability. For the first time geography had ceased to be a barrier, and the U.S. mainland lay exposed to enemy fire. In that respect, Russia's rockets were infinitely more frightening than the Japanese bombers that had attacked Pearl Harbor sixteen years before. It was not distant naval bases on Pacific islands that they targeted, but the impregnable heartland itself: Cincinnati, St. Louis, Chicago, Detroit, places that had never before needed to worry about foreign aggression. Despite White House assurances to the contrary, satellites and ballistic missiles were inherently linked. The story, therefore, was ultimately about the security—or newfound insecurity—of the American people, as Sevareid made plainly clear: "If the intercontinental missile is, indeed, the ultimate, the final weapon of warfare," he ended his broadcast ominously, "then at the present rate, Russia will soon come to a period during which she can stand astride the world, its military master."

The warning was echoed by thousands of media outlets, big and small, conservative and liberal, in radio and television, magazines and newspapers. Sputnik was "a great national emergency," declared Max Ascoli of the *Reporter*. A "grave defeat," lamented the staunchly Republican *New York Herald Tribune*. *US News & World Report* likened it to the splitting of the atom. The editors of *Life* made comparisons to the shots fired at Lexington and Concord and urged Americans "to respond as the Minutemen had done then." Sputnik was "a technological Pearl Harbor," fretted Edward Teller, the father of the H-bomb. The sphere's "chilling beeps," echoed *Time*, were a signal that "in vital sectors of the technology race, the US may have well lost its precious lead."

A strange sense of disconnection gripped the public discourse. The more the administration told Americans not to worry, the louder the media beat their doomsday drums. Editors seemed obsessed with the Soviet satellite, and pretty soon so was the general population, which had initially greeted the launch with mild to complete disinterest. "The reaction here indicates massive indifference," a *Newsweek* correspondent had reported from Boston on October 5. "There is a vague feeling that we have stepped into a new era, but people aren't discussing it the way they are football or the Asiatic flu," another *Newsweek* reporter wired from Denver. In Milwaukee, it was the ballistic trajectories of the Braves' pitching staff in Game 2 of the World Series against the Yankees that preoccupied most people, not the short news brief on page 3 of the *Sentinel* devoted to the Soviet satellite. According to a spot poll conducted on October 5 by the Opinion Research Corporation, only 13 percent of Americans saw Sputnik as a sign that America had fallen dangerously behind the Soviet Union. One reason so few people were worried, recalled the Columbia University pollster Samuel Lubell, was an overwhelming sense of confidence in Eisenhower's leadership. "When I asked what this country should do, the reply would fairly often be: 'The President will do all that needs to be done,'" Lubell noted. "Or, a typical answer would be: 'He's taking action now.' Or 'I'd leave that to the President. He ought to know.'"

Within days the media barrage changed the public mood dramatically. People began holding nightly vigils to try to spot the passing satellite; they tuned their radios to its frequencies; and they grew anxious. Yet for the Democrats in Congress, Sputnik was simply too good an opportunity to let slip. The Little Rock crisis had left Eisenhower vulnerable, and the economy was weakening. The Soviet Union had handed the United States a setback that could be whipped up into a full-blown indictment of the administration.

As the most vocal critic of Eisenhower's "deplorable" military cuts, Symington took the lead, rallying his fellow Democratic hawks Henry "Scoop" Jackson of Washington and Richard Russell of Georgia, the powerful chairman of the Senate Armed Services Committee. The launch was proof, Symington said, "of growing Communist superiority in the all-important missile field." The administration's "penury" had let America's technological lead slip away and had placed the nation in grave danger.

"I have been warning about this growing danger for a long time," Symington added, "because the future of the United States may well be at stake." He asked Russell to convene hearings immediately so that "the American people [can] learn the truth."

Naturally, Symington volunteered to lead the investigation, as he had during the bomber gap. Both Russell and Jackson had been around Washington long enough to know that he had ulterior motives, but they were only too happy to oblige their handsome and ambitious young colleague. He had credibility, and the *New York Times* had praised his poise and his "dignified bearing that conveys an impression of statesmanship."

Jackson enthusiastically took up the cause, calling for "a National Week of Shame and Danger." Sputnik, he said, was "a devastating blow to the prestige of the United States." Russell weighed in as well. "We now know beyond a doubt," he warned on October 5, "that the Russians have the ultimate weapon—a long-range missile capable of delivering atomic and hydrogen explosives across continents and oceans. If this now known superiority over the United States develops into supremacy, the position of the free world will be critical. At the same time we continue to learn of the missile accomplishments of the possible enemy. For fiscal reasons this Government, in turn, continues to cut back and slow down its own missile program."

But Symington was not the only ambitious politician looking to capitalize on the Communist feat. Lyndon Johnson had been at his Texas ranch on the night of October 4, when news of the Soviet satellite had reached him. Like Eisenhower, he loved his rural retreat and "liked nothing better than to career over the hills in his convertible Lincoln Continental, shooting bucks from the front seat," in the words of Johnson's biographer Randall B. Woods. On Sputnik night he had been entertaining guests at his deer tower, an air-conditioned, glass-enclosed, forty-foot-high hunting blind, complete with a dining room and a staff of black waiters. It sat at the wooded edge of a meadow and was flanked by banks of powerful spotlights that Johnson would switch on, blinding his prey for an easy shot. But that night, he had laid down his rifle and drinks and stared at the sky. "I'll be dammed," he swore, "if I sleep by the light of a Red Moon."

He would also be damned if he was going to let Symington grab the spotlight. He was the majority leader, the most powerful legislator in the land, and if he ever hoped to be taken seriously as a presidential contender he needed to weigh in on the crisis. "Soon they will be dropping bombs on us from space like kids dropping rocks onto cars from freeway overpasses," cried Johnson, who until then had never expressed particular interest in either missile technology or space, adding his own calls for "a full and exhaustive inquiry" into the sorry state of national defense.

Johnson rushed back to Washington and began plotting. His first order of business was to head off Symington, and his first call was to Richard Russell, the man who would decide which Armed Services subcommittee would hold the Sputnik hearings. Russell was Johnson's ace in the hole. The quiet and courtly senator from Georgia was a model of old-fashioned southern gentility and probably the most powerful Democrat in Washington—"the undisputed leader of the Senate's inner Club," in the words of the historian Doris Kearns Goodwin. Johnson, from the first day he set foot in the Senate, recognized Russell's immense influence and systematically sought to curry his approval. Always addressing him respectfully, without any of the jocular familiarity he reserved for other lawmakers, Johnson bombarded Russell with polite notes and queries that made it clear he valued his opinions. He took care to be on hand on Saturdays and late evenings, when Russell, a bachelor with no outside social life, was alone in the empty Senate. "I made sure that there was always one companion, one Senator, who worked as long as and as hard as he, and that was me," Johnson later recalled. Johnson also made a point of getting himself appointed to Russell's Armed Services Committee to cement the burgeoning relationship. "I knew there was only one way to see Russell everyday," he explained, "and that was to get a seat on his committee."

In time, Lady Bird Johnson began extending invitations for the lonely, workaholic Georgian to join the Johnsons for Sunday brunches and holiday meals, and by 1957 a special bond had been forged between the two senators. That relationship would now come in handy.

As rival Democrats battled over who could ring the alarm bells loudest, the orchestrated histrionics had their desired effect. Public reaction to Sputnik quickly shifted from blasé to terror-stricken. Everywhere around the country, people flocked to rooftops and held midnight vigils on their front lawns, hoping to catch a glimpse of the ominous orb whose signal was being blamed for a rash of mysterious garage door openings. (The *Washington Post* speculated that these were the result of interference from coded messages to Soviet spies.) Local radio stations fueled the paranoia by broadcasting Sputnik's expected over-pass times, and it was not unusual to have entire blocks of people gazing anxiously skyward at 3:00 AM. Eventually, some 4 percent of the U.S. population would report seeing Sputnik with their own eyes. (What most actually saw was the one-hundred-foot-long R-7 rocket casing that Korolev had craftily outfitted with reflective prisms. It trailed some 600 miles behind the twenty-two-inch satellite, which could be viewed only with optical devices more sophisticated than the binoculars used by the average American.)

Throughout the ruckus, Eisenhower remained resolutely silent. If the administration stayed calm, he was certain, the furor would pass. People would see that the sky was not falling in on their heads and would return to their normal routines. "Business as usual" was the message the

White House chose to project. "We can't always go changing our program in reaction to everything the Russians do," Eisenhower told his cabinet. But as two, three, and then four days passed without any public comment from the president, the press grew irritable and impatient. Everyone else, it seemed, had pronounced on the subject of Sputnik; where did Ike stand? Was America really in danger? Did the Soviets in fact possess the ultimate weapon of mass destruction? Where was the leadership? Newspapers, especially in the South, which was still seething over Little Rock, demanded to know. "Ike Plays Golf, Hears the News," grumbled the *Birmingham News*, while the *Nashville Tennessean* ran a cartoon of the president dismissing Sputnik from the putting green.

Privately, Eisenhower's aides were anything but dismissive, and there was growing concern that Ike's purposeful silence was backfiring. "This was a place where Eisenhower went wrong," his loyal staff secretary General Andrew J. Goodpaster conceded decades later. "His expression was that this was nothing we didn't foresee or know about, but the American people until that moment had not realized the vulnerability that had now developed. That they could be reached by long range rockets, which could be nuclear armed. And our country, for the first time, was exposed to that kind of danger. And so, where he brushed it off as something that we had foreseen, it really created great anxiety, almost panic within the United States."

Eisenhower's background as a professional soldier may have been partly responsible for his empathy deficit. As a military man, the president was accustomed to calculating casualties and collateral damage. From his experiences in World War II, he knew that in modern combat there was no longer any such thing as noncombatants; the United States had long targeted Russian cities, and it was not that shocking that the Soviet Union did the same. As a seasoned field commander, Eisenhower also knew that the ICBM, as a weapon, was still in its infancy, much like the airplane before World War I, and that years would pass before it became a real threat that could alter the balance of power.

The president and the military men who served in his immediate circle were not attuned to the psychological effects of Sputnik as a symbol of nuclear Armageddon. "I can't understand," Eisenhower told Goodpaster, "why the American people have got so worked up over this thing. It's certainly not going to drop on their heads."

Others in the administration, however, were better equipped to appreciate the national trauma. Vice President Nixon, as a career politician with limited military experience, instinctively grasped that Sputnik could not be shrugged off lightly as a "stunt." It was a mistake, he argued privately (and later in his memoirs), not to acknowledge it as a serious affront to American supremacy; and he would be the first senior administration official to say so publicly, during a speech in San Francisco on October 15. The White House press secretary Jim Hagerty was also deeply worried by the media onslaught Sputnik had generated. His boss was taking a lot of flak and needed to devise a strategy to disarm his critics. Especially troublesome was the negative publicity being stirred up by an Associated Press story that the army had been prevented from launching a satellite in 1956. The leak apparently infuriated Eisenhower and was the subject of a damage-control session he held with his military and science advisers at 8:30 AM on Tuesday, October 8. Donald Quarles took the brunt of Eisenhower's anger. "There was no doubt," Quarles admitted, "that the Redstone, had it been used, could have placed a satellite in orbit many months ago," but he was quick to spread the blame, adding that "the [Pentagon] Science Advisory Committee had felt that it was better to have the earth satellite proceed separately from military development. One reason was to stress the peaceful character of the effort."

Ike was not pleased. "When this information reaches the Congress," he observed, frowning, "they are bound to ask questions."

Eisenhower may have come to politics late in life, but he was hardly naive enough to hope that the Democrats would not try to pin the blame on him. And Quarles, as Charlie Wilson's unenthusiastic point man on satellites, had been the ranking administration official responsible for turning down Medaris's repeated requests to convert the Redstone-based Jupiter C into a launch vehicle. The deputy defense secretary, though, tried to put a positive spin on the potential public relations disaster. "The Russians," he said in a feeble attempt to make the news seem welcoming, "have in fact done us a good turn, unintentionally, in establishing the concept of freedom of international space."

Eisenhower knew he could not tell the American people that there was a silver lining to the Soviet breakthrough—that the United States would be able to phase out the secret U-2 overflights and spy on the

USSR from space without violating international laws. Still, he had to say something to mollify the public, and at Hagerty's urging he finally agreed to hold a press conference the next day.

· · ·

When Eisenhower walked into conference room 474 of the Old Executive Office Building at precisely 10:31 AM on Wednesday, October 9, he was greeted by one of the most hostile press corps the president had ever faced. Hagerty, anticipating angry questions about why the army had not been permitted to use a loaded orbital stage during the Jupiter C trials, had prepared a two-page statement that was distributed shortly before the president's arrival. "The rocketry employed by our Naval Research Laboratory for launching our Vanguard," it explained, "has been deliberately separated from our ballistic missile efforts in order, first, to accent the scientific purposes of the satellite and, second, to avoid interference with top priority missile programs. Merging of this scientific effort with military programs could have produced an orbiting United States satellite before now, but to the detriment of scientific goals and military progress. Our satellite program," the statement concluded, "has never been conducted as a race with other countries."

The White House press corps was not pleased. "Mr. President," demanded Merriman Smith of United Press International, "Russia has launched an earth satellite. They also claim to have had a successful firing of an intercontinental ballistic missile, none of which this country has. I ask you sir, what are you going do about it?"

Eisenhower was not accustomed to this sort of treatment, and he appeared surprised by the ferocity of the question. Photographs show him scowling, eyebrows arched, leaning across the microphone, pale in a dark tie and charcoal three-piece suit. The president had always enjoyed a friendly and jocular relationship with the men and women who covered him, and often played a game of making his press conferences as obtuse and unintelligible as possible to avoid delicate topics. This time, though, the assembled journalists were in no mood for meandering answers.

Eisenhower delivered a lengthy response to Smith's question that reiterated America's intention to put up a satellite as part of its IGY efforts but offered little concrete evidence of a new plan of action or any juicy sound bites. Charles von Freed of CBS was not satisfied. "Mr. President," he said, "Khrushchev claims we are now entering a period when conventional planes, bombers and fighters will be confined to museums because they are outmoded by the missiles which Russia claims she has perfected. Khrushchev's remarks would seem to indicate he wants us to believe our Strategic Air Command is now outmoded. Do you believe that SAC is outmoded?"

"No," Eisenhower shot back emphatically. The process, he explained, would be evolutionary rather than revolutionary, and would take twenty years.

May Craig of the *Portland Press Herald* kept up the pressure. "Mr. President, you have spoken of the scientific aspects of the satellite. Do you think it has immense significance in surveillance of other countries?"

"Not at this time," Eisenhower obfuscated, not mentioning that just the day before he had grilled Donald Quarles on the progress of the air force's lagging space reconnaissance program. "I think that period is a long ways off when you consider that even now, and apparently they have, the Russians, under a dictatorial society, where they have some of the finest scientists in the world, who have for many years been working on it, apparently from what they say they have put one small ball in the air."

This was the sound bite that everyone had been waiting for. Eisenhower's dismissive "one small ball" would grace hundreds of headlines in the next day's newspapers, reinforcing the impression that the president of the United States was at a loss as to why his nation was so traumatized. Pleased, the reporters pressed on. "Mr. President," the *Chicago Tribune* correspondent queried, smelling blood, "considering what we know about Russia's progress in the field of missiles, are you satisfied with our own progress in that field, or do you feel there have been unnecessary delays in our development of missiles?"

This was precisely the type of loaded question that Richard Nixon had predicted during the NSC meeting two years before, when he had argued with Quarles that the administration had to be seen as doing everything in its power to move forward with the new weapons systems. Eisenhower had not attended that meeting, and now he seemed hesitant. "I can't say there has been unnecessary delay. I know that from time to time I came here and got into the thing earnestly," the president started to say, but then abruptly changed tack. "We have done everything I can think of . . . I can say this: I wish we were further ahead and knew more as to the accuracy and to the erosion and to the heat resistant qualities of metals and all the other things we have to know about. I wish we knew more about it at this moment."

"Is it a correct interpretation of what you said about your satisfaction with the missile program as separate from the satellite program," the *Washington Post* reporter followed up, "that you have no plans to take any steps to combine the various government units which are involved in this program and which give certainly the public appearance of a great deal of service rivalry, with some reason to feel that this is why we seem to be lagging behind the Soviets?"

"First of all, I didn't say I was satisfied," Eisenhower replied testily. "I said I don't know what we could have done better."

More probing questions followed in the same cutthroat vein. Why did Charlie Wilson, the day before, on his last day in office, say he doubted the Soviet Union had an ICBM? Did the sudden cancellation of a state visit by Soviet Marshal Georgy Zhukov have anything to with Sputnik? Was it true that the army was being prevented from launching a satellite immediately? Would the United States launch a satellite as heavy as Sputnik?

Finally, NBC's Hazel Markel cut to the chase: "Mr. President, in light of the great faith which the American people have in your military knowledge and leadership, are you saying at this time that with the Russian satellite whirling about the world, you are not more concerned nor overly concerned about our nation's security?"

Eisenhower's measured response delved into the difficulties of missile accuracy and the still relatively primitive state of guidance systems. But the details were lost on the journalists. His answer would be pared down on the evening news to a single flippant sentence fragment: "Not one iota."

• • •

If the president's news conference had been intended to pacify the press, it had the opposite effect. Instead of reassuring the public with his trademark calm and commanding demeanor, Ike's performance was judged to have been too remote, too divorced from the anxiety sweeping the nation. "A fumbling apologia," snipped one critic. "A Crisis in Leadership," declared *Time*, noting that American voters wanted a strong leader, unfazed by crisis. But they also needed someone to understand and address their fears. By dismissing Sputnik as "a small ball" without military implications, the man Americans trusted most to defend them seemed oblivious to the danger that millions now saw lurking in the night sky. Ike was correct that in itself the Soviet satellite posed no danger, but he failed to acknowledge that it represented a potential threat. Instead of projecting confidence, he was accused of being out of touch with reality, asleep at the wheel. The president must "be in some kind of partial retirement," complained the hugely influential syndicated columnist Walter Lippmann. "He is not leading the country," said the usually supportive *New York Times* columnist Arthur Krock. The *Washington Evening Star* was even less charitable, comparing Eisenhower's subdued reaction to that of someone under the effects of mind-numbing sedatives. Only the conservative *US News & World Report* rallied behind the president, calling his refusal to be cowed "courageous statesmanship."

In Congress, delighted Democrats heaped scorn on the administration's refusal to recognize or respond to the Soviet challenge. They accused Eisenhower of "penny-pinching" on missiles, of "complacency" on satellites, "lack of vision" on both, and "incredible stupidity," in general. Republican lawmakers, sensing the shifting tide in public opinion, toned down their defense of the White House and prepared to weather the storm. As a Republican congressman confessed, "No greater opportunity will ever be present for a Democratic Congress to harass a Republican administration, and everyone involved on either side knows it."

No one on Capitol Hill was more keenly aware of this than Lyndon Johnson. His chief strategist, a wily former wire-service reporter by the name of George Reedy, penned out the possibilities. "The issue of [Sputnik], if properly handled," Reedy wrote in a sweeping memo, "would blast the Republicans out of the water, unify the Democratic Party, and elect you President." Sputnik, he added, was just the ticket the Democrats needed to supplant integration as the major campaign issue in the upcoming midterm and presidential campaigns. The disgrace of Little Rock and the continued opposition by Dixiecrats to integration were threatening to split the party and were likely to prove costly at the polls. Sputnik, argued Reedy, presented a unique opportunity "to find another issue, which is even more potent. Otherwise the Democratic future is bleak."

National security was that issue, and now Johnson began maneuvering to place himself at its center, to batter Eisenhower, and to elbow Symington aside. Johnson was no stranger to matters of defense. He had first vaulted to national prominence during the 1950 hearings by the Senate Armed Services Subcommittee on Preparedness, which had investigated America's missteps early in the Korean conflict. That inquest had established the Texan as a man to watch, who skillfully deflected Republican criticism of President Truman. That Truman himself had ridden the chairmanship of the very same subcommittee during World War II to the vice presidential nomination and ultimately to the presidency had not been lost on Johnson, who now, seven years later, began lobbying his mentor Richard Russell to convene Preparedness hearings instead of holding an Air Power inquest.

Johnson started working his legislative magic, calling in favors and swapping promises. Hushed conferences were held in corridors and cloakrooms; telephone calls were placed to the right people. Elbows were squeezed; shoulders were patted. Winks and nods were exchanged on the Senate floor, as deals were brokered over power breakfasts and intimate dinners at the Mayflower Hotel. It was the famous Johnson Treatment, and no one could withstand it for long. "Its velocity was breathtaking," in the description of Rowland Evans and Robert Novak. "Its tone could be supplication, accusation, cajolery, exuberance, scorn, tears, complaint, the hint of threat. He moved in close, his face a scant millimeter from his target, his eyes widening and narrowing, his eyebrows rising and falling. From his pockets poured clippings, memos, and statistics. Mimicry, humor, and the genius analogy made 'The Treatment' an almost hypnotic experience and rendered the target stunned and helpless."

The majority leader had no rivals when it came to bending the Senate to his will. Senator Hubert Humphrey of Minnesota called Johnson's elaborately scripted seductions an art form, "making cowboy love." And Russell now found himself at the receiving end of Johnson's persistent affections. Johnson argued that as the chairman of the Preparedness subcommittee, he would be best positioned to steer the national discourse away from unpopular Democratic positions on civil rights. Symington, he argued, was interested only in a hysterical witch hunt. He, on the other hand, would take the high road, adopt a more tactful approach that on the surface appeared bipartisan and patriotic, but which would be just as devastating.

At the White House, an alarmed Vice President Nixon was keeping tabs on the scheming Texan, the man he well knew was lining up to challenge him in 1960. Go to Congress, he urged Ike, to defuse the situation. Disarm Johnson and Symington by offering concessions that will knock the wind out of their sails. Bring the Senate into the conversation before it tries to bring you down, he counseled, knowing that it was his political future that could suffer most in the long run. But Ike, who had a stubborn streak that belied his gentle, outwardly docile nature, declined. He would maintain a posture of business-as-usual and would not dignify the uproar by pandering to a bunch of self-serving lawmakers.

At an emergency session of the National Security Council on Thursday, October 10, Nixon once more advocated a stronger response. The session was opened by Allen Dulles, who outlined Sputnik's far-reaching implications. "We do not, as of yet, know if the satellite is sending out encoded messages," he said. "Furthermore, we must expect additional launchings."

Dulles had warned Eisenhower in late September that a Soviet launch was imminent, and he and Nixon had proposed going public with the information to lessen the potential shock. Their suggestion had been rejected, and now Dulles listed the consequences. "Khrushchev has moved all his propaganda guns in place," he said. Sputnik was merely "one of a trilogy" of public relations coups. "The other two being the announcement of the successful testing of an ICBM, and the recent test of a large scale hydrogen bomb at Novaya Zemlya. Incidentally," he added, the Soviets had just exploded another big H-bomb "late last night."

The close timing of the three feats, Dulles noted, was having "a very wide and deep impact" abroad. The Chinese, he said, were treating Sputnik as proof of Soviet military and technological supremacy over the United States. Similar statements were coming out of Egypt and other countries in the Middle East, and Moscow was giving "the theme maximum play" with its Eastern European satellite states. Even America's Western European allies, Dulles reported, were rattled, and confidence in NATO, particularly in France, had taken a psychological hit. All in all, the situation on the foreign policy front was "pretty somber."

Eisenhower interrupted at one point to inquire about Sputnik's weight: Was it really so heavy? he asked, saying he had heard that "someone here had gotten a decimal point out of place." Unfortunately, he was assured, it was indeed 184 pounds.

The floor was then turned over to Quarles, who once more launched into an impassioned defense of why the government had chosen to separate the IGY project from ballistic missile programs to pave the way for spy satellites. "In this respect," Quarles reiterated his point from Tuesday's meeting, "the Soviets have now proved very helpful. Their satellite has over-flown practically every nation, and thus far there have been no protests."

Two things stood out about the Sputnik launch, Quarles reluctantly conceded. First, it was "clear evidence that the Soviets possess a competence in long-range rocket and auxiliary fields which is more advanced than we had credited them with." And second, the outer space reconnaissance implications of the launch were "of very great significance."

Both conclusions contradicted Ike's public statements. But they must have rung loud and clear for Allen Dulles, for already Richard Bissell had approached him with the idea of quietly stealing the spy satellite mandate away from Quarles and the air force, as they had done with the U-2.

The conversation turned to Vanguard's progress and then returned to the political fallout from Sputnik. Arthur Larson, Eisenhower's chief speechwriter, cleared his throat hesitantly, as if he had something unpleasant on his mind. "I wonder if our plans for the next great breakthrough are adequate," he finally said, referring to the decision not to change the IGY plans or to get ABMA involved in the satellite race. "If we lose repeatedly to the Russians as we have lost with the earth satellite, the accumulated damage will be tremendous. We should accordingly plan, ourselves, to achieve the next big breakthrough first, a manned satellite," he suggested, "or getting to the moon."

Nixon rallied behind Larson. The vice president was among the few administration figures "who seemed to grasp the new symbolism," in the words of the historian Walter McDougall, and once more he advocated announcing immediate increases in both missile and space spending. "The country will support it," he said. But Eisenhower cut the discussion short. Everyone around the table, he warned, would soon be called to testify before congressional committees. There could be no dissension in the ranks. Everyone had "to stand firmly" behind the decision to stay with the current course.

"In short," concluded the president, "we should answer queries by stating that we have a plan—a good plan—and we are going to stick to it."

• • •

Plan? fumed Medaris. What plan? He and von Braun had been following the developments in Washington with increasing fury and apprehension. The political storm that von Braun had predicted had indeed materialized. But the administration's response was a far cry from what Medaris had anticipated when he ordered ABMA to begin satellite preparations without authorization. The official go-ahead had never come, and as the days passed without word from Washington, it was becoming increasingly clear that he had jumped the gun.

Jim Hagerty, meanwhile, apparently also jumped the gun, announcing that the Vanguard program would launch "a small satellite sphere" in December, before sending up a fully instrumented scientific payload in March 1958. The dates were according to the original IGY schedule. But the December launch had been intended as a quiet dress rehearsal to check Vanguard's booster and upper stages, which had never been fired in tandem before and still bore the technical designation "experimental vehicle" to denote their untested status. The press secretary, however, had made the date of the test public, guaranteeing that the whole world would be watching what was never meant to be anything more than a trial run. The dress rehearsal had effectively become opening night. "We who could coldly appraise the odds of Vanguard were frankly scared to death," Medaris recalled, somewhat insincerely. Virtually every new rocket system failed on its first attempt, as both Wernher von Braun and Sergei Korolev could attest from bitter experience. That was why inaugural flights were kept quiet. Only after the kinks were ironed out were successes reported to the general public. Thanks to the

White House, however, Vanguard would debut on a national stage, live, in front of television cameras. It was a recipe for disaster, Medaris believed. "How far out on a limb could our poor country get?" he said.

For his part, Medaris was also getting "far, far out on a limb" with his rogue satellite preparations. For the first few days it had been easy to hide the unauthorized expenditures, but as one week turned into two, and then two weeks became three, the costs were mounting, and it was becoming impossible to bury the paperwork. Von Braun's team had even started eating into the newly reinstated overtime budgets. (The administration, fearing the coming congressional inquests, had quietly rescinded its ban on overtime for missile projects.) Medaris was putting his own career on the line. "I had neither money nor authority, yet work was still going on," he later confessed. "By the end of the month, I was really sweating, and beginning to wake up in the middle of the night talking to myself."

As October edged toward November, pushing Medaris farther out on his limb, only two things could happen, he felt increasingly certain. Either Vanguard would fail, in which case the government would have no choice but to turn to ABMA as its last resort. Or he was going to face a full court-martial, a dishonorable discharge, and possibly prison. It never occurred to him that Nikita Khrushchev might provide a third option.

9

SOMETHING FOR THE HOLIDAYS

Dwight Eisenhower wasn't the only one caught off guard by Sputnik. Nikita Khrushchev had also initially underestimated its hefty political payload.

Before October 4, Khrushchev had been only partly paying attention to the proceedings at Tyura-Tam. "Just another Korolev launch," he later conceded, recalling that an aide had needed to remind him that the Chief Designer, in a fit of paranoia, had moved up the date by two days. Since the R-7 had already proved itself on two successful trials, there was no longer any great sense of urgency as to the rocket's viability. Its temporary incarnation as a space launcher, while intriguing from a scientific and competitive point of view, was not critical to the missile's main mission as a weapon. The stakes, therefore, were not so high, at least as far as the first secretary was concerned.

Khrushchev also had some pressing earthly problems to contend with. Like Korolev, he had fallen prey to paranoia and fear of rivals, both real and imaginary. In the weeks that followed the summer's failed hard-liner putsch, he had become increasingly convinced that another coup was in the works, and that once more dark forces were aligning to depose him. His nagging doubts festered, so that by the time Korolev rolled out the R-7 for his space shot, Khrushchev had decided to act on his suspicions. But he had to tread cautiously and spring the subtlest of traps, because his perceived challenger this time was not a party rival or a Stalinist holdover but the head of the Soviet armed forces, a soldier with the nation at his feet and the world's largest army at his command.

Marshal Georgy Zhukov, hero of the Great Patriotic War, conqueror of Berlin, savior of Moscow, and Khrushchev's rescuer during the June coup, had simply grown too powerful. The man whose popularity had so intimidated Joseph Stalin that the old tyrant had not dared have him killed was once more impinging on the balance of power within the Kremlin. Nor, it seemed, could he help himself; he was just too large for life, and he kept threatening to overshadow his civilian masters.

To ordinary Russians, Zhukov was a legend, a Soviet Patton and MacArthur rolled into one deliciously gruff and outsize package. Arrogant and abrasive, he had a soldier's disdain for politicians, a seaman's penchant for profanity, and a marine's storm-the-beaches attitude toward bureaucracy. Like Khrushchev, he had been born poor and humble from illiterate peasant stock, and he had spent a childhood laboring in factories instead of classrooms. His real education had come on the battlefield, starting at the age of seventeen. Marked for early promotion in the new Red Army because of his proletarian roots and daring cavalry charges during the October Revolution and the civil war that followed, Zhukov had pioneered the use of tanks on his rapid rise. It was those innovative tank tactics that had first brought him to Stalin's attention in July 1939. In one of his infamous fits of paranoia, Stalin had just butchered forty thousand officers, including most of his general staff, and Japan had exploited the resulting vacuum and disarray in the Soviet High Command to seize Mongolia, an unofficial Soviet dominion. Dispatched to repel the invaders, Zhukov routed the Japanese so soundly that they sued for peace and signed a nonaggression treaty that in time would protect the USSR from having to fight on two fronts. Impressed, Stalin summoned the young general to Moscow in 1940 to lead the German side in war games that simulated a Nazi invasion. Once more, Zhukov routed his adversary, seizing the Kremlin; an alarmed Stalin appointed him chief of the general staff, responsible for preparing for a real war with Germany, which the Soviet leader refused to believe would ever come. When it did, as Zhukov and many others had predicted, the marshal resigned rather than accept Stalin's stubborn belief that the line had to be held at Kiev. Zhukov had wanted to fall back farther east and regroup, laying the same trap that had lured Napoleon in 1812 and forcing the enemy to stretch its supply lines in the dead of the Russian winter. When Hitler smashed through to Moscow's outskirts in a matter of months, Stalin—in a rare and uncharacteristic moment of humility—begged his insubordinate general to return, even naming him deputy commander in chief, a de facto admission that Russia's fate was now in his hands. Zhukov drove the invaders back all the way to Hitler's underground bunker in Berlin. But victory came at a high cost: Zhukov lost more than one million men in the battle of Stalingrad alone, and his tactics were said to be so brutal and callous that they seemed premised on the notion that his enemies would run out of bullets before he ran out of soldiers to send to the slaughter.

Some of his officers hated him for the unnecessary carnage, but the people loved him for rescuing the nation. "Where you find Zhukov, you find Victory," a saying was coined, and after the war, Stalin was so jealous of Zhukov's status as the most decorated soldier in Soviet history that he had him removed from his exalted perch. Yet even the murderous Stalin was too afraid of a backlash to arrest or execute Mother Russia's favorite son. Zhukov was merely sent to rot away in a series of meaningless posts far from the capital until Stalin's death, when Khrushchev rehabilitated him to legitimize his coup against Beria.

Khrushchev had also invoked Zhukov's popularity and unimpeachable reputation as a national patriot to stare down Lazar Kaganovich and the other coup plotters when they had the upper hand. And of course it was Zhukov who ferried, via long-range bomber, the Central Committee members whom Khrushchev had needed for his own survival, and it was the marshal who delivered the main charges against the conspirators at the extraordinary plenum that had been held after the putsch.

Zhukov had emerged from the failed coup as a kingmaker, arguably the second most powerful man in Russia—and probably the country's most revered public figure, for he did not carry with him the taint of Communist Party purges. He was a product of the military, historically the nation's most trusted institution, especially after its glorious role in the Great Patriotic War. Alas, the military was now unhappy with Khrushchev, particularly for his ruinous love affair with missiles.

Since starting work on the ICBM, Khrushchev had unilaterally slashed troop forces by a staggering 2 million men. He had canceled long-range bomber orders and converted aircraft factories to building passenger planes. Military airfields had been turned over to civilian use, under the expectation that rockets would soon arrive to protect the Soviet Union. Entire artillery divisions had been similarly scrapped, while dumbfounded admirals had helplessly watched brand-new battle cruisers—"shark fodder," as Khrushchev derisively called them—get cut up into scrap steel before ever having a chance to leave their naval shipyards. Following the R-7's August success, Khrushchev had announced a further round of three-hundred-thousand-troop reductions for the end of the year. Only submarines, which would carry the unproven missiles on which he was basing his entire defense doctrine, saw an increase in orders.

Not surprisingly, the wholesale cuts had roiled the Soviet armed forces. The uniformed services were not at all convinced that missiles were the panacea Khrushchev was promising, and they resented having to give up their heavy artillery, battle tanks, cruisers, and infantry units to pay for the experiment. "Some voices of dissatisfaction were heard blaming me for this policy," Khrushchev later recalled. Resentment rippled through the ranks as entire commands were lost with the closures of bases, battalions, air wings, and naval squadrons. Thousands of careers had suddenly ground to a halt, and disgruntled murmurs could be heard at officers' clubs from East Berlin to Sakhalin Island. The massive personnel cuts had not yet sliced too deeply into the elite officer corps, but the writing was on the wall. Without troops to command, the majors and colonels, and even some generals, would soon lose their jobs, along with the prestige and, more important, the perks—the access to better food and housing, cars and drivers—that went with their privileged positions.

The growing discontent within the middle and upper echelons of the military was all the more troubling since Khrushchev was uncertain whether he could count on the general staff's loyalty. Foolishly, he had permitted Zhukov to pension off most of the old military leaders that the Soviet leader had known for years. "I can't go to battle with generals who have to travel with field hospitals," Zhukov had explained of the need for fresh blood in the geriatric High Command. Khrushchev had consented, but now he regretted his decision. He was not personally acquainted with many of the new crop of Red Army leaders that had been promoted in their place, and this younger generation of officers owed its allegiance directly to the charismatic defense chief. More ominously, after being elevated to full voting membership of the Presidium for his role in foiling the putsch, the marshal was now also free to build tactical alliances with Khrushchev's Communist Party colleagues, further expanding his potential power base. "He assumed so much power that it began to worry the leadership" Khrushchev later observed in his memoir, though by "leadership" he of course meant himself.

Politically, Zhukov appeared to have outfoxed his insecure master, who soon became convinced that the marshal coveted "Eisenhower's Crown"—to be both head of the military and head of state. "Father feared that Zhukov saw General Eisenhower as an example," Sergei Khrushchev recalled. "I see what Zhukov is up to," Khrushchev told his fellow Presidium members in late September. "We were heading for a military *coup d'état*."

The clincher appeared to have been a particularly unsettling piece of intelligence that Khrushchev received a few weeks earlier. His defense chief, he was told, had secretly started "saboteur schools" outside Moscow and Kiev to train highly specialized covert operations teams. Typically, the Central Committee was informed whenever new military units were created, but Zhukov had not followed protocol and had kept his civilian overseers ignorant of his activities. Only the new head of military intelligence, a confidant whom Zhukov had recently appointed, had been kept in the loop. There could be only one explanation for the lapse, Khrushchev reasoned; the urban commandos Zhukov and his spy chief were secretly training were setting the stage for "a South American–style military takeover."

Whether the conspiracy was real or whether it was a figment of Khrushchev's inflamed imagination would become a matter of historical debate. (Not even Sergei, with the benefit of hindsight and a half century's distance, would be able to unequivocally support his father's suspicions.) That Khrushchev, after his brush with insurgency, had become more prone to seeing potential plots was perhaps understandable. That Zhukov had grown far more powerful than any other soldier in the Soviet era was also undeniable. But had he actually harbored mutinous ambitions? And had he really wanted to rule Russia? On that score, the historical jury is still out, and even Khrushchev would later wonder if he had jumped the gun.

Zhukov's lapse in reporting the saboteur schools might have been an innocent omission, or simply a pretext for Khrushchev to launch a pre-emptive strike against an ally whose growing influence was becoming too dangerous. Whichever the case, by late September he had decided to remove Zhukov before the defense chief could remove him. "His unreasonable activities leave us no choice," Khrushchev told his loyalists in hushed meetings. The only remaining question was how to do it without tipping his hand and risking a military revolt. Zhukov could not be approached head-on. He needed to be isolated, cut off from contact with his subordinates, and lulled into a false sense of security. And all this had to be accomplished so subtly that he would never see the ax coming.

The answer lay in foreign travel. Zhukov had been invited to Washington. Why not go to Yugoslavia instead, Khrushchev suggested. Andrei Gromyko, the new foreign minister, would be in Washington anyway, and the uppity Yugoslav leader Josip Broz Tito needed some hand-holding ahead of the big pan-Communist summit in Moscow that was coming up in a few weeks to celebrate the fortieth anniversary of the October Revolution. Zhukov could take one of the brand-new battle cruisers (one of the few that had survived the scrap heap), it was further suggested, to impress the Yugoslavs, and he could visit Albania as well.

And so, on October 4, the unsuspecting defense minister was somewhere in the Adriatic, stuck on a slow boat from Tirana, his radio communications restricted, while Khrushchev hastened to Kiev to seal his fate. Ostensibly, the Ukrainian stopover was routine, a chance for the boss to meet with regional party officials, listen to their petitions for funds, and discuss economic policies. Somewhat less routinely, Khrushchev was also there to observe tank maneuvers and to meet with senior officers from the Kiev Military District, one of Zhukov's former commands. Select members of the brass from Moscow had flown in for the impromptu talks, most notably Rodion Malinovsky, the deputy defense minister, whom Khrushchev still trusted.

Whether the timing of the visit with military commanders was purely coincidental would also become a topic of historical speculation. No record was kept of what was said during the maneuvers. A few days later, however, a small and unobtrusive squib would appear on the back page of *Pravda*. Marshal of the Soviet Union Georgy Konstantinovich Zhukov, it would announce in the smallest of print, had been "relieved of his duties." Rodion Malinovsky was assuming his responsibilities.

On the night of October 4, the matter of Zhukov's removal was either not yet resolved or still a closely held secret, because Sergei Khrushchev, who had joined his father in Kiev, had no inkling of the preemptive countercoup. Sergei had seen little of his father over the past several months. He had just gotten married, was busy with his dissertation, and had hopes of soon going to work for Korolev. He was beginning to make the transition from favored son to an adult with his own independent life and family, vacationing separately for the first time that summer and no longer living at home. Apparently Nikita missed Sergei's company, because he had called him the day before, suggesting he hop on a plane and meet him at the Marinsky Palace in Kiev, where Khrushchev would be making an unexpected detour on his way to Moscow from his seaside dacha in the Crimea.

The palace crowned Kiev's highest hill, offering postcard views of the Dnieper River and the gilded, green domes of the one-thousand-year-old Pecherskaya Lavra Monastery on one side, and, on the other, the more foreboding sight of the new gray granite government buildings built by German prisoners of war. Inside, beneath pendulous chandeliers, pastel ceilings, and ornate millwork, Sergei sat, bored, as he waited for his father to finish his seemingly endless meetings. The evening sessions had run well past ten, in the customarily rambling fashion of official Soviet delegations, so that the air in the Marinsky Palace dining hall had grown stale with cigarette smoke, and the crystal ashtrays around the long dining table were beginning to overflow. Among the smokers were Khrushchev's closest supporters, and it was probably also not a coincidence that he had chosen to be with the chieftains from his former power base—he had served as Stalin's Ukrainian viceroy—on the day he was orchestrating the removal of the man he judged to be his final rival.

In the palace's convivial atmosphere, Khrushchev could cast a reassuring and proprietary eye on men whose careers he owned: the suddenly ever-present Rodion Malinovsky; Aleksei Kirichenko, the powerful Ukrainian party boss who had accompanied Khrushchev to Korolev's design bureau in February 1956, when the Chief Designer had requested permission to launch a satellite; Leonid Brezhnev, the dim-witted but trustworthy loyalist, who was actually neither but found it expedient to play the part. Khrushchev had just put Brezhnev in charge of missile and defense matters at the Presidium, a reward for his (literally) swooning support during the botched coup. The post seemed largely ceremonial, since the first secretary still made all the important decisions himself. But it sent a signal that the bushy-haired young political commissar from Dnipropetrovsk was on the rise; and, just as important, that loyalty paid off.

It was Khrushchev, not Brezhnev, who shortly after 11:00 PM was summoned away from the meeting on a missile-related matter. An aide, Sergei Khrushchev recalled, whispered in his father's ear that he had a phone call. "I'll be back," Khrushchev announced, leaving the room. He returned a few moments later, a broad smile creasing his tired features. He said nothing, though, and for some time sat silently, staring at his fingernails in a distracted manner as he listened to reports on the beet harvest and coal stockpiles for the coming winter. After a few minutes of fidgeting, however, he could no longer restrain himself and raised his hand for silence.

"Comrades," he said, addressing the assembled Ukrainian Central Committee members. "I can tell you some very pleasant and important news. Korolev just called." (At this point Khrushchev acquired what his son would later describe as "a secretive look.") "He's one of our missile designers. Remember not to mention his name—it's classified. So," Khrushchev continued, "Korolev has just reported that today, a little while ago, an artificial satellite of the Earth was launched."

The Ukrainians stared blankly, not quite sure what to make of this news. Obviously, the boss was pleased, but about what few could tell. Most people in the room had never heard of a satellite before. "Everyone smiled politely, without understanding what had just happened," Sergei recalled. Khrushchev, perhaps sensing the perplexity of his provincial underlings, felt compelled to explain. "It is an offshoot of an intercontinental missile," he said.

This additional intelligence did not appear to make the visibly confused local party bosses any wiser, so once more the Soviet leader elaborated. As Sergei described the scene:

> Father began talking about missiles. He spoke of how the appearance of ballistic missiles had radically altered the balance of forces in the world. His audience listened in silence. They seemed completely immersed in his account, but their faces revealed their indifference. They were used to listening to Father, regardless of the subject. The Kiev officials were hearing about missiles for the first time and clearly didn't understand what they were.

Missile doctrine might have been above the pay grade of the regional party hacks but, ironically, it was Khrushchev who had missed the point. Despite his unusually prescient grasp of rocketry's role in modern warfare, he had completely failed to recognize Sputnik's significance as a propaganda weapon. "He had viewed the satellite primarily in military terms," Sergei conceded. For the Soviet leader, Sputnik had been a milestone in the ICBM race, not a milestone in human history. That mankind had just broken the bounds of gravity, and made its greatest leap—to quote CBS's Eric Sevareid—from the primordial mud, was completely lost on the missile-obsessed Khrushchev. Which may have explained why the lead story in *Pravda* on the morning of October 5 was incongruously titled "Preparations for Winter" and tallied food and fuel stockpiles for the coming cold season, while Sputnik was relegated to a terse two-paragraph news brief.

"We still hadn't realized what we had done," Sergei Khrushchev remembered. It would take the world to tell them.

• • •

For Sergei Korolev, it would also take a while for the magnitude of his accomplishment to sink in. There had been little time to celebrate after the launch: a few rounds of vodka and congratulatory speeches in the middle of the night, followed by intensive calculations and worries as the first day of the space age wore on. PS-1, as everyone at Tyura-Tam still called the satellite, was up, but how stable was its orbit? This the anxious scientists could not immediately determine, because their tracking stations were arrayed only on Soviet territory and thus could measure only a small fraction of the satellite's elliptical orbit. Sputnik would have to make at least a dozen full revolutions before anyone could tell with certainty whether it would stay in orbit or come crashing back to earth.

Every celestial body suffers from what is known as orbital decay, a gradual loss of speed and altitude that brings the object either closer to or farther from its center of gravity. Decay can be imperceptibly slow, as in the case of the moon, which falls a few inches away from the earth every year, or catastrophically abrupt, like a meteor getting sucked in by the pull of gravity. Sputnik could thus stay in orbit for a day, a week, a month, a year, a millennium, or a million years, depending on that all-important rate of degeneration.

From Sputnik's first few rotations, Korolev's team had been able to ascertain the satellite's basic parameters: its apogee, perigee, speed, inclination, and duration of each orbit. The tiny sphere was hurtling on a 25,000-kilometer-per-hour (15,625-miles-per-hour) roller-coaster ride around the planet, crossing the equator every ninety-six minutes at a sixty-five-degree angle as it climbed to apogees (peaks) of 947 kilometers (587 miles) above sea level like a surfer on a wave and then plummeted to perigees (depths) of 228 kilometers (141 miles) as it fell to the bottom of a trough before rising again. But each time Sputnik hit a trough was like slamming on the brakes, because the atmosphere, even at that height, was still thick enough to cause friction. At such relatively low points of the orbit, fluctuations in the earth's gravity due to differences in the shape and mineral composition of the globe, which is not a perfectly round sphere, could also adversely affect decay. Sputnik's perigee was too low because of the malfunction during liftoff, which had resulted in an early engine cutoff. The question was, How much ground was it losing? If PS-1 fell back to earth within a few days, Korolev's triumph would be short-lived, his record tainted, his masters in Moscow unhappy. Frantically, his mathematicians ran the numbers, trying to predict Sputnik's life span. Finally, early in the afternoon of October 5, they came up with an estimate: two to three months. (The exact number would turn out to be ninety-two days.) Everyone breathed a sigh of relief. Sputnik was safe, as far as the record books and politicians stood. Korolev and his chief designers could at last relax and go home to celebrate in earnest.

Until then, there had been no time for reflection. "We were all too focused on our jobs, concentrating on the execution of the operation, to think about the meaning of the event," recalled Vladimir Barmin, the designer of the Tulip launchpad. Boris Chertok described feeling a similar sensation of relieved exhaustion rather than euphoric wonder on finally hearing that the space barrier was broken. "It was late. We went to bed," he wrote, with uncharacteristic brevity, of getting the news at OKB-1 headquarters in Moscow, where he had been recovering from his illness. "We thought the satellite was just a simple device," he added, "and that the importance of the launch had been to test the R-7 again and gather data."

Of all the engineers, physicists, chemists, mathematicians, and military personnel who had been involved with Sputnik, literally several thousand people, it seemed that only the erudite Mikhail Tikhonravov, the Latin-spouting creator of PS-1, understood that the world had changed forever on October 4, 1957. "This date," he said, "has become one of the most glorious in the history of humanity."

The men responsible for the satellite would begin to grasp the importance of their feat only when they boarded their special flight from Tyura-Tam to Moscow on the night following the launch. Most of the exhausted engineers had passed out shortly after takeoff, Valentin Glushko and Mstislav Keldysh slumbering in their elegant and neatly pressed suits, while Korolev, shifting uncomfortably in his trademark black leather jacket and turtleneck sweater, stared wearily at the dim cabin lights. As soon as their big Iliushin-4 prop jet had leveled off over the orange Kazakh desert, the pilot, Tolya Yesenin, came rushing out of the cockpit. "The whole world is abuzz," he gushed, grasping the Chief Designer's hand and pumping it furiously. Korolev sat up, startled. He had been so preoccupied during the past twenty-four hours that he had had little contact with anyone outside Tyura-Tam other than Khrushchev, and had no idea that word of the launch had spread so far, so wide, so fast. Abuzz? The whole world? Really? Korolev couldn't contain himself. He jumped out of his seat and made straight for the flight deck to use the plane's radio. When he returned some minutes later, he was unusually ebullient and emotional.

"Comrades," he cried, rousing his sleepy colleagues. "You can't imagine what's happening. The whole world is talking about our little satellite. Apparently we have caused quite a stir."

• • •

Much like Korolev, it was not until the night of October 5, and only once he had returned to Moscow from his maneuverings in Kiev, that Nikita Khrushchev began to realize what a tremendous victory he had just scored against the United States.

Throughout the day, Soviet embassies and KGB stations around the globe had been busy compiling foreign press clippings and political reactions to Sputnik. By the next morning, the reports had been translated, cabled to Moscow, sorted, and slotted into the thick folders Khrushchev received with breakfast every day at his government mansion in Lenin Hills. The files—green for foreign press clippings, red for decoded diplomatic traffic, blue for agency reports—must have made savory reading. "The achievement is immense," declared Britain's *Manchester Guardian*. "It demands a psychological adjustment on our part towards Soviet society, Soviet military capabilities, and perhaps—most of all—to the relationship of the world to what is beyond. The Russians can now build ballistic missiles capable of striking any chosen target anywhere in the world. Clearly they have established a great lead in missile technology." "Myth has become reality," crowed France's *Le Figaro*, commenting gleefully on the bitter "disillusion and bitter reflections of the Americans who have little experience with humiliation in the technical domain."

Khrushchev leafed through the stack of diplomatic dispatches with increasing relish. "A turning point in civilization," the *New York Times* declared, "that could only be achieved by a country with first rate conditions in a vast area of science and engineering." An Austrian paper opined that "in contrast with the first steps in the atomic age which began with 100,000 deaths, mankind can rejoice without destruction on the conquest of cosmos by the human spirit." China's main daily hailed Sputnik as a "validation of the superiority of Marxist-Leninist technology." Radio Cairo declared that "the planetary era rings the death knell of colonialism; the American policy of encirclement of the Soviet Union has pitifully failed."

Khrushchev was astonished by the reaction. It was as if, overnight, his nation had been vaulted to a preeminent position atop the global hierarchy. The Soviet Union, in the eyes of the world, had suddenly become a genuine superpower, not just a backward and brutish empire to be feared because of its sheer size, territorial ambitions, and aggressive ideology but a true and equal rival of the United States, a beacon of progress that deserved respect for its technological prowess and forward thinking. "With only a ball of metal," as the historian Asif A. Siddiqi would succinctly put it, "the Soviets had managed to achieve what they were unable to convey with decades of rhetoric."

The turnaround, the KGB reported, had profoundly shaken America's allies in both tangible and esoteric ways. The European Assembly in Strasbourg censured the United States for falling behind the Soviet Union. In Tehran, the shah's CIA-sponsored government "considered the satellite such a blow to U.S. prestige," according to a diplomatic assessment, "that they displayed uneasy embarrassment in discussing it with Americans." In Mexico, editors had begun requesting Soviet rather than U.S. scientific source material, while Japan's ruling Liberal Democratic Party had taken Sputnik as a cue to begin "agitating" against further deployment of U.S. conventional armed forces. As the U.S. Information Agency would itself concede, in an October 10 memo, "Public opinion in friendly countries shows decided concern over the possibility that the balance of military power has shifted or may soon shift in favor of the USSR. American prestige is viewed as having sustained a severe blow, and the American [domestic] reaction, so sharply marked by concern, discomfiture, and intense interest, has itself increased the disquiet of friendly countries and increased the impact of the satellite."

The cold war had suddenly taken on a new and, from Khrushchev's perspective, eminently more appealing dimension. It was now the specter of Soviet supremacy rather than American dominance that haunted the global arms race. Moscow, for once, held the high moral ground in this new phase of the contest because Sputnik, as opposed to Hiroshima, could be touted as a purely peaceful and scientific achievement. That must have been the most delicious irony for Khrushchev. He had tried, and failed, to rattle the world with his announcement in August boasting of a deadly new weapon that would raise the scale of mass destruction to unprecedented levels. People had simply shrugged. Korolev, on the other hand, had placed a tiny transmitter on top of an R-7 and managed to put the entire planet on notice with its innocuous little beeps. "It will generate myth, legend, and enduring superstition of a kind peculiarly difficult to eradicate," the USIA memo accurately predicted, "which the USSR can exploit to its advantage."

The Soviet leader smiled. He had read enough. He may have lacked the formal education and erudition to intuitively grasp the historic context of man's ascent to the heavens, but he was too well grounded a politician not to recognize opportunity when it knocked. "People all over the world are pointing to the satellite," he exclaimed, as if struck by a revelation. "They are saying the U.S. has been beaten." Pushing aside his breakfast, and all previous thought of the dearly deposed Zhukov, Khrushchev sprang into action. *Get me Korolev!* he ordered.

· · ·

By the time the Chief Designer arrived at Khrushchev's Kremlin office on Thursday, October 10, the propaganda apparatus of the Communist Party of the Union of Socialist Soviet Republics had been marshaled and unleashed for the benefit of its citizenry. *Pravda* and every other Soviet mass medium now single-mindedly pursued the glorious twin topics of space and satellites, trumpeting the proud and limitless promise of Soviet science. "World's First Artificial Satellite of the Earth Created in Soviet Union," *Pravda*'s October 6 issue hailed in a page-wide banner headline. "Russians Won the Competition," the paper boasted the next day, in equally bold print. Western newspaper articles from publications usually reviled as corrupt capitalist organs were duly reprinted on front pages to reassure the Soviet citizenry that the world thought as much of Sputnik as the ITAR-TASS news agency claimed. Laudatory telegrams from leaders of various fraternal nations were published to show the awed reverence of the Communist bloc. Photos and footage of the racism in Little Rock were once more disseminated to drive home the stark contrast between Soviet innovation and American oppression. The campaign was relentless, and it was just as effective as the media blitz in the United States that had turned audience indifference to terror—only in reverse. Ordinary Russians, who days earlier had neither heard of nor cared about satellites, were suddenly space converts overwhelmed with national pride and a newfound sense of security and superiority.

The name Sergei Pavlovich Korolev was notably absent from the triumphant barrage of state television, radio, and print reports. Nor were the names Glushko, Pilyugin, Barmin, or any of the other top designers and engineers ever mentioned in the same breath as PS-1, which by then had also acquired the generic moniker Sputnik with Russian audiences. The public face of the Soviet space program in fact belonged to a man who had nothing to do with rockets or satellites. Leonid Sedov, a technocrat and expert in gas dynamics, was the Academy of Sciences' representative at the IGY and various other international conferences. His chief recommendation as Soviet spokesman appeared to be a fluency in English and German, a polished delivery, and a talent for obfuscation. Sedov had become something of a celebrity since October 4, particularly in the West, where his snide remarks on America's moral decline made all the newsreels. "The average American only cares for his car, house, and electric refrigerator," he lectured during one speech. "He has no sense of national purpose, nor is he receptive to great ideas which do not pay off immediately." Meeting Ernst Stuhlinger after a speech at the Eighth Congress of the International Astronomical Federation in Barcelona, Spain, on October 8, Sedov was less flip but just as smug. "We could never understand," he lectured, according to the security memorandum Stuhlinger filed with ABMA's counterintelligence bureau detailing his contact with the Communist official, "why your people picked such a strange design [Vanguard] for a satellite carrier. Why did you try to build something entirely new instead of using one of your excellent military engines? You would have saved so much time, not to mention troubles, and money. Why did Dr. von Braun select this other design?"

Stuhlinger appeared both puzzled and frustrated at the mistake in identity. "Dr. von Braun?" he replied to Sedov. "He did not decide this. He is not a member of the Vanguard Committee; in fact he is not even a consultant or adviser on the American Vanguard satellite." Sedov no doubt relayed this nugget of intelligence in his own debriefing report to the KGB, along with reassurances that his German-American interlocutors had never inquired about Korolev, whose secret identity was presumably still safe. It was unlikely that Sedov himself would have inadvertently passed on any classified information. Famous as he was becoming as the spokesman for Russian rocket science, he apparently knew so little about the actual workings of missiles that on a visit to Tyura-Tam, he astounded Korolev by asking where the satellites were placed on the R-7's central booster. (On top, Korolev had dryly replied.)

If the Chief Designer resented talking heads like Sedov stealing his limelight, he did not say so. "People in the Soviet Union did not complain during that era," Sergei Khrushchev laughed when asked if Korolev found the enforced anonymity grating. Korelev's daughter, Natalia, however, recalled her bitter disappointment when the Nobel committee wanted the name of the scientist responsible for Sputnik so they could award him the Nobel Prize in Physics. It is the collective achievement of Soviet science, the Swedes were told. "I remember walking in Red Square," Natalia Koroleva recounted decades later, "and seeing all these banners, and celebrations, and I wanted to shout, 'My father did this.' But I couldn't tell anyone."

"They are well provided for," Khrushchev said of his nameless missile experts. Their identities, he regretted, had to remain secret for national security reasons. But one day, he vowed, "we shall erect a monument in honor of those who created the rocket and Sputnik and shall inscribe their glorious names in letters of gold so that they will be known to future generations."

For now, they would have to make do with medals they could not wear publicly. In addition to the Order of Lenin and Hero of Socialist Labor awarded to all those involved with PS-1, Korolev received an honorary doctorate and was elevated from corresponding to full member of the Soviet Academy of Sciences. Those accolades, however, were fairly meaningless. His chief reward came with his October 10 summons to the Kremlin—and the visit's tacit recognition of his admission into Khrushchev's inner circle of court favorites. "Our most brilliant missile designer," Khrushchev raved, noting that other rocket designers "could not hold a candle to Sergei Pavlovich Korolev." That status would confer on the Chief Designer a rare and privileged position, and the ability to cut through red tape and circumvent bureaucratic hurdles on his future ventures. Henceforth, he would have a direct line to the Soviet Union's sole decision maker and could bypass the sort of annoying obstacles that the R-7 State Commission had thrown in his way prior to Sputnik's launch.

While Korolev knew he could not take public credit for Sputnik, he was hardly blind to the considerable political capital he had earned with Khrushchev. The Soviet leader was making so much hay from his satellite that he would be hard-pressed to deny him any reasonable requests—and Korolev was not the sort of person to shy away from pressing his advantage. Neither, of course, was Khrushchev, which made them an ideally suited pair.

The first secretary might have been slow recognizing the propaganda value of the hand Korolev had dealt him, but once he had belatedly realized what kind of cards he held, he had wasted no time capitalizing on his windfall. He had already summoned James Reston, the *New York Times* bureau chief, to his Kremlin office so he could communicate directly to the American people. "When we announced the successful testing of an intercontinental rocket," Khrushchev told Reston, "some American statesmen did not believe us." The Soviet leader was referring to Charlie Wilson's famously dismissive comments on the R-7, perhaps even to Eisenhower's veiled skepticism. "The Soviet Union, they claimed, was saying it had something it did not really have," Khrushchev went on, revealing a glimpse of his bruised ego. "Now that we have successfully launched an earth satellite, only technically ignorant people can doubt this."

The Americans had also laughed, Khrushchev continued in this wounded vein, when the Soviet Union had announced its intention to launch a satellite. Sputnik was up, he chortled, and where was the American satellite—the one the size of a grapefruit? "If necessary, we can double the weight of the satellite," he boasted, adding that the ICBM it rode on was "fully perfected" and could strike anywhere in the world. What's more, Khrushchev vowed, growing overly animated as he often did when discussing sensitive subjects, the R-7 would soon go into mass production, and ICBMs would roll out of Russian factories "like sausages."

The tirade, duly relayed in all its frightening implications to alarmed American readers, had not been just an outburst of pent-up frustration or even a manifestation of Khrushchev's notorious inferiority complex. It had been a coldly calculated feint, a bluff designed to deflect attention away from the fact the Soviets were discovering that the R-7 had serious limitations as a so-called ultimate weapon. "Initially, Father believed the mere existence of an ICBM would deter the Americans," his son explained. But the assumptions behind that doctrine had been shattered during the R-7 tests, when Marshal Nedelin and the military discovered how much time it took to fuel the huge rocket and how difficult it was to hide. War, it was now thought, could still break out because the missile was vulnerable to preemptive strikes and could be destroyed on the launchpad by U.S. bombers, rendering its deterrent value nil. Washington, after all, already knew Tyura-Tam's exact geographic coordinates thanks to its spy planes, and the R-7 could be launched only from that one location because it was too big to be moved on anything other than railcars and needed a pad the size of several football fields from which to lift off. It also used the wrong kind of propellant, not to mention a staggeringly impractical 250 tons of it, requiring cumbersome fueling infrastructure and hours of wasteful preparation time because its tanks could not be prefilled with liquid oxygen that instantly evaporated. The R-7, in short, could not be hidden, moved, or fired on short notice—making it a sitting duck in the event of a surprise American attack.

As with many first-generation weapons, its principal value was in its demonstrative effect; the dream of deterrence via an invulnerable ICBM fleet was realistic, if not yet a reality. Already one of Korolev's rival designers, Mikhail Yangel, was working on a series of successor missiles that addressed the R-7's flaws. Yangel, a few months earlier, had successfully tested the R-12, an intermediate range missile that used storable nitric acid—Glushko's preferred oxidizer—instead of slow-loading liquid oxygen. With Glushko's backing, Yangel was now proposing an expanded intercontinental version of the missile, the R-16, which would be a third the size of the original R-7 and capable of silo or mobile launch on less than thirty minutes' notice. Glushko, who could not help but envy the political accolades that the Chief Designer was garnering for what was essentially a triumph of his own engines, was lobbying Nedelin to push for the R-16. It would take three to four years for the superior weapon to go from blueprint to deployment phase, but the Americans did not know any of this. For now, at least, the R-7 was still the only ICBM in existence on either side of the ideological divide. And Nikita Khrushchev was not about to let anyone forget that he, and he alone, had exclusive domain over the power to rain destruction anywhere on the planet.

The point was not just that the Soviet Union possessed this devastating new weapon, but that Khrushchev wielded it personally. The R-7 was his creation—Korolev was merely an instrument of his will—and it was to him that the political accolades ultimately fell, a point that *Pravda* would hammer home whenever possible. "In his able proposals," the paper would glowingly note, "there is evidence again and again of the great conviction in the triumph of Soviet rocket technology." Likewise, Sputnik celebrated his glory ("he participates in the discussions of all the most vital experiments") and validated his vision (he "directs the development of the major directions of technical progress in the country"); he would invoke the satellite in virtually every speech for months to come. For a leader still deeply insecure about his own authority, Sputnik was a boon. It was the glue that Khrushchev had been looking for to cement his grip on power. Everyone knew that it was his rocket up there causing an international sensation. Everyone would associate his name with one of the greatest technological triumphs of the twentieth century. With luck, he might even ascend to that rarefied pantheon of Russia's Greats: Peter, Catherine, Stalin. They had been immortalized for their terrestrial conquests; Khrushchev had just expanded that empire into outer space.

For the embattled Soviet leader, October 4 had augured an astonishing reversal of political fortune: his final rival had been eliminated, forgotten in the furor over Sputnik's success, and he alone had emerged as the prime beneficiary of the satellite's conferred grandeur. Khrushchev could now claim credit for making the Soviet Union a genuine superpower, a true technological match for the United States. On the world stage, he was now Eisenhower's equal. At home, he was untouchable, safe at last from intrigue, schemers, and prospective coup plotters. But still, Khrushchev wanted more. Sputnik could be squeezed for even greater political gains, and he could scale even higher political heights. Korolev, he was sure, could make it happen.

· · ·

"You know," said Khrushchev, when the Chief Designer finally arrived at his Kremlin office on the morning of October 10, "when you first proposed Sputnik we didn't believe you. We thought, 'Ah, that Korolev, he's just dreaming.' But today it's another story."

Khrushchev beamed at his star scientist. He was in exceptionally high spirits. The boss, as Korolev knew, had been cranky of late and easily flew off the cuff. But that morning he seemed completely relaxed, lounging in a sofa chair next to fellow Presidium member Anastas Mikoyan. The wily old Armenian, a survivor of Stalin's inner circle who owed his political longevity to an utter lack of ambition, also struck an informal pose, nibbling at a bowl of fruit while he stretched his plump legs on an expensive central Asian rug. Tea and juice had been offered along with the easy banter, a sign that Korolev, who had dressed for the occasion in a respectful tie and jacket, was in unusually good standing.

"Sergei Pavlovich," Khrushchev continued, "as you know, the October Revolution jubilee is approaching." Korolev needed no reminder. He would have had to have been blind not to notice the frenzied preparations for the fortieth anniversary of the Bolshevik uprising. Moscow's main streets were getting a makeover in anticipation of the parades and ceremonies that would mark four decades of communism, and construction crews were busy repaving roads, repainting buildings, and scrubbing soot from grimy facades. Sputnik would be a major theme of the celebration, a symbol of Soviet accomplishment; Khrushchev had commissioned poems lyricizing the "Leap Forward" and had ordered that detailed timetables be published in every major city showing exactly when Sputnik passed over different metropolitan regions. Huge Sputnik banners were erected, commemorative stamps printed—FIRST IN SPACE, they boasted—and millions of spherical, satellite-shaped pins were made for citizens to wear on their lapels. They would bear them proudly because Sputnik had tapped an unusually sensitive nerve with ordinary Russians. Most Soviet citizens knew that life was different in the West, that people in Europe and America enjoyed higher standards of living under capitalism. In that respect, Khrushchev's inferiority complex was a national malaise. Sputnik compensated for those persistent feelings of inadequacy and inequality. Muscovites might not have color television sets, fast cars, or fashionable shoes, but Sputnik proved that they weren't technologically backward after all. " 'Now we are ahead of America,' I have been told countless times," reported Tom Margerison, a British science writer on assignment in Moscow, in the

London *Sunday Times*. "In the streets there is immense pleasure and pride in the rocket-engineer's achievement. . . . In Red Square I counted no fewer than fourteen models of Sputnik circling a globe. . . . Their success is more important to the Russians themselves than to anyone else."

Sputnik had given Moscow the high moral ground over the West, demonstrating how shallow consumerism should be sacrificed for the good of science and human progress. While it was presented as a monumental triumph, Sputnik in reality helped cover up and justify one of the most glaring shortcomings of communism: its inability to deliver basic material well-being to its citizens. As Margerison acidly put it, "Nowhere else would you find a people who are able to carry out a complex project like launching a satellite, involving the close cooperation of scientists and engineers from many disciplines, yet who prove quite unable to organize efficient butcher shops." As a substitute for comfort, and as a tool to pacify the masses, now that terror had been rejected, Sputnik was thus invaluable. It exploited pride rather than fear, and it supplanted Stalin's purges as a way of keeping people committed to wobbling socialist ideals.

All this must have flashed through Khrushchev's head as he prodded Korolev. "It'll be forty years of Soviet power, which is a big milestone. Wouldn't it be nice," the Soviet leader asked wistfully, "to have something for the holiday?"

The question, with all its implications, hung in the air momentarily. But it was obvious where Khrushchev was headed. He had invited all the leaders of world communism to Moscow for the anniversary (which because of the switch from the old Julian czarist calendar would actually take place on November 7) and wanted another feat to impress his honored guests. He was especially keen to woo China's Mao Zedong with Korolev's magic, since the Chinese had been growing increasingly aloof and independent ever since his secret speech at the Twentieth Party Congress. Beijing had blasted Khrushchev's assault on Stalin as "revisionist" and had made ugly noises about no longer recognizing Russia's role as the ideological standard-bearer for communism. But Mao was veritably smitten with missiles and had openly marveled at Sputnik. Khrushchev, eager to bring the Chinese back into Moscow's fold, had promised Mao missile technology, starting with the R-2, which was to be transferred in 1958. (The Chinese, in turn, would transfer the technology to their client states, and in time the R-2's DNA would figure in virtually all future generations of Asian missiles.)

Khrushchev clearly relished the prospect of Mao in Moscow, salivating at another display of Soviet missile muscle flexing. Mikoyan also seemed to have caught the gist of his boss's loaded question. "Maybe," he suggested, "a Sputnik that will broadcast the 'Internationale' from space." The "Internationale" was the pan-Communist anthem, a hymn that reverberated for the October Revolution in much the same way as the "Marseillaise" sounded the French Revolution.

But Khrushchev didn't think much of the idea. "What?" he snapped, cutting off a cowed Mikoyan. "You and your Internationale. Forget the Internationale. [Sputnik]'s not a damn music box."

Glaring briefly at Mikoyan, Khrushchev once more turned to his Chief Designer, his expression immediately softening, his eyes attentive and hopeful. Korolev's mind must also have been racing throughout the exchange. He had the parts to assemble one more rocket; otherwise the next batch of R-7s would not be ready until January 1958. So he had a launch vehicle. But what about a satellite? Nothing was ready on that front, and he couldn't simply replicate PS-1. Whatever he came up with had to top the original Sputnik: be bigger, better, and create an even greater sensation. The bar was significantly raised, and the time frame was beyond brutal. It had taken three years to launch Sputnik. Khrushchev was giving him barely three weeks.

A wiser man might have said that it was impossible, that it couldn't be done. But Korolev did not hesitate. "What if we launch a Sputnik with a living being?" he asked nonchalantly, as if building a spacecraft from scratch in a matter of days was the easiest thing in the world. The military, Korolev explained, had been sending dogs on high-altitude, suborbital rocket flights and parachuting them back to earth in special hermetically sealed compartments. He could borrow one of those canine chambers, outfit it with a life-support system, and cobble something together.

Khrushchev's face lit up as he listened. "With a dog in it!" he exclaimed, as if the idea had just spontaneously come to him and Korolev was an extension of his own iron will, merely a mechanic who filled in the blanks and fussed over the details. "Can you imagine, Anastas?"

Khrushchev cried triumphantly, addressing Mikoyan, who was nodding enthusiastically, as he always did whenever the first secretary looked to him for reassurance. "A dog in space."

It would be such a coup. Not only could the USSR stake another claim to cosmic supremacy, but it would also be the first nation on the planet to prove that life could be sustained beyond earth's boundaries. Once more, the whole world would stand in awe of Soviet science while trembling at the strength of its missiles. It was perfect.

"That's what we need," a clearly animated Khrushchev continued. "A dog. Give us a dog. But," his features darkened in a vaguely ominous, finger-wagging way, "make sure you are ready for the holidays."

It was Korolev who now nodded with forced sincerity, since he had little choice. "We will do our best, Nikita Sergeevich," he promised, sounding somewhat less certain, being sure to use the collective *we* in case blame later had to be spread.

"We are agreed then, Sergei Pavlovich," Khrushchev said, standing up to indicate that the interview was over. "You will have whatever you need. You can ask my man Kozlov"—another new Politburo appointee—"for whatever you want. Meet with him tomorrow to go over the details. But remember," Khrushchev admonished, "we need this for the holidays."

• • •

Twenty-six days. The number must have reverberated in Korolev's head like an oppressive drumbeat, like the pounding of a migraine that could not be dulled. What had he gotten himself into? Twenty-six days to design, build, test, and launch a spacecraft. Scratch the testing; there would be no time for that. The design phase would also have to be severely curtailed if he had any hope of meeting his deadline. There would be no special drawings. His engineers would have to make crude sketches and give them directly to machinists, to be produced without quality control. But what about the overall concept? How big would the craft have to be to keep an animal alive? And for how long? They would have to feed their canine cosmonaut remotely, monitor its progress electronically, process its waste hydraulically, and provide it with a steady supply of fresh oxygen. That was a lot of equipment to haul into orbit. Would Glushko's engines carry the extra weight? What about the heat? How would they shield their passenger from the forces of friction and solar rays? Would the capsule require a special shroud? And how would it separate from the launch vehicle once it had reached orbital velocity? Ejection systems were complex and prone to malfunction. They couldn't risk one, not without extensive testing. Perhaps they could weld the satellite to the R-7's core booster and try to blast the whole thing into space. But that solution presented its own problems.

Dilemmas and technical conundrums swirled through Korolev's mind as his black limousine pulled out of the Kremlin gates. The Chief Designer was not particularly introspective or prone to soul-searching panic attacks. But he could not have failed to wonder whether his ambition and supreme self-confidence had exceeded his better judgment on this occasion.

Even if Mikhail Tikhonravov, his resident satellite expert—the "chief theoretician of cosmonautics," as he would soon be called—could sketch a few rudimentary blueprints, they would have to build PS-2 on the fly, improvise virtually every step. The hardware would have to come entirely off the shelf, since there was no time for new components or anything fancy. Korolev hated cutting corners. But he hadn't left himself a choice.

Nor would there be room for error. This time Khrushchev would be watching, and the entire Presidium would be expecting results. This launch was pure politics, and that was always dangerous ground. If Korolev was successful, he could write his own ticket. And he had ambitious plans to cash in his political chips.

The Chief Designer had not idly proposed orbiting a dog with PS-2. He had not wildly blurted out the suggestion. It had been premeditated, a prelude to his ultimate goal: to plant the seeds for a space program that would someday put human beings in orbit and ultimately a man on the moon. Officially, the Soviet Union had no such designs. Satellites, to Khrushchev, were offshoots of missiles. Korolev wanted to change that perception and make space a politically viable destination in its own right, a propaganda weapon. PS-1 had opened Khrushchev's eyes to that possibility. This second Sputnik could seal the deal.

If he failed, though, the space option would forever be off the table.

Khrushchev would not give him a second chance. His interplanetary dreams would be over, and he would spend the rest of his career working on ICBMs—if he was lucky. Everything depended on PS-2.

Korolev's first order of business was to get his team back to Moscow OKB-1 headquarters, and that in itself was no easy task. After Sputnik's launch, he had given all his top engineers time off to recuperate from the months of heavy exertion at Tyura-Tam. He himself could have used a break. He was exhausted, having worked himself to the point of collapse. But he'd stuck around Moscow, expecting the call from Khrushchev, knowing that the opportunity that might present itself had to be seized. Now he could only hope that his health would hold up another few weeks, and that his men could get back on time from various resorts on the Baltic and Black seas. The Chief Designer started issuing frantic recall notices. "My wife and I were in Kudespa on the Baltic when I received the telegram from Korolev to return to Moscow immediately," Evgeny Shabarov recalled. "I went to the airport but couldn't get any tickets." Soviet airlines were always booked months in advance, especially from tourist destinations. "I went to see the airport administrator, and showed him the cable," Shabarov went on. "Oh yes, I know all about you, he said, here's your ticket." Korolev had called the transport ministry, warning its officials that Khrushchev would have their heads if every single one of his rocket scientists was not back the next day.

"We're returning to Tyura-Tam tomorrow," the Chief Designer announced when all his astonished engineers had been assembled. "Be prepared to go back to work."

10

OPERATION CONFIDENCE

"**S**oviets Orbit Second Artificial Moon; Communist Dog in Space," screamed the headlines on the morning of Monday, November 4, 1957, as Americans awoke to another media riot and fresh rounds of recriminations.

"What next?" demanded the *New York Herald Tribune* incredulously. "A Man on the Moon?" "Moscow Mission to Mars in Near Future?" the *Washington Star* speculated, its editorial dripping with defeat and resignation. "Shoot the Moon, Ike," urged the feistier *Pittsburgh Press*, suggesting defiantly that the White House blow the offending Soviet satellite to smithereens.

From his desk on the second story of the Old Senate Office Building, Lyndon Johnson surveyed the stack of alarmist articles, the barest hint of a smile creasing his craggy features. "Plunge heavily into this one," advised an accompanying note from his aide George Reedy. But Johnson needed no exhortations from underlings to spot the opportunity he had been waiting for. For several weeks now he'd been sitting on the sidelines, gauging the political winds. The deals had been cut, and his rival, Stuart Symington, had been dispensed with. The only remaining question had been the timing, since Congress was not in session.

During the past month, Johnson had busily poisoned the well against Symington, who may have "looked most like a President," in the opinion of the *New York Times*, but had proven no match for the master of the "Johnson Treatment." The Democratic majority leader had not only wooed Richard Russell but also sweet-talked the ranking Republican members of the Armed Services Committee, Styles Bridges and Leverett Saltonstall. "Let's not look for scapegoats," he told them, "but let's find out what's wrong and let's do what's necessary to fix it." Johnson conceded that Symington was the Senate's leading expert on missiles, but he said mournfully that the Missourian was also out for blood, looking only to hold Air Power hearings that would be "too shrill" in tone, "too partisan" in nature. The Preparedness subcommittee, on the other hand, would be bipartisan and interested only in "solving the problems," not apportioning blame. Never mind that the subcommittee was defunct and had not been used in years, while Symington's investigative body was fully staffed and an ongoing concern. Johnson sealed the deal by promising Bridges and Saltonstall that they could help preside over the Preparedness inquest, knowing full well that the men were up for reelection in 1958.

The matter of the subcommittee had thus been settled. Johnson, however, had not rushed out and announced his intentions to hold Preparedness hearings. For all his reputation as a freewheeling horse trader, as a reckless and charmingly relentless rogue, he was an inherently cautious legislator, never putting himself out front of an issue unless the outcome was guaranteed. Like the lawyer who only asked questions for which he already knew the answer, Johnson only supported measures to which he'd already secured prior passage. He knew that challenging President Eisenhower on national security, no matter how subtly, was a risky proposition. But as Johnson scanned the hysterical articles on Sputnik II, the satellite's specifications made him bolt upright. The thing was monstrous: a staggering 1,120 pounds, and well over three terrifying tons when the rocket casing to which it was welded was factored into the equation. Unlike its predecessor, this second Sputnik was almost as heavy as a hydrogen bomb—incontrovertible proof that the Soviet Union did in fact have the capability to hurl heavy nuclear warheads at the United States, despite Eisenhower's dogged assurances to the contrary.

Johnson knew he now had his ammunition, the silver bullet he and Senator Russell had been waiting for to take on the popular president.

"Sputnik II absolutely made the decision for them," recalled his aide Glenn P. Wilson, "because it weighed so much more."

The following day, Johnson and Russell called a press conference on the steps of the Pentagon. The members of the media, by then, had already worked themselves up into a speculative frenzy over Sputnik II's passenger, the mix-breed terrier Laika, and her planetary laps. "The greatly increased size of the second Sputnik means that it was probably not fired by the same rocket system that launched the first one," *Time* opined erroneously. "This is enough weight allowance to put a powerful atomic bomb on the moon," the magazine added, also erroneously. The *New York Times* fared no better, wondering "whether the Soviet Union might be using some new form of rocket propellant unknown in the West" (which it was not) to generate so much lift. Analysts at the Pentagon and the CIA, meanwhile, feverishly revised their estimates of the Russian ICBM's thrust from a too-low 500,000 pounds to a too-high 1.5 million pounds—still out of range of *Time*'s moon shot but more than enough to plunk a hydrogen bomb anywhere on earth.

Johnson and Russell whipped up the overeager press. Pronouncing themselves deeply "alarmed" at the briefing they had just received from the Joint Chiefs of Staff on the relative state of American missiles, they announced that an emergency session of the Preparedness subcommittee would be convened later that month. "As Chairman of the Committee, the Senate Democratic leader reported that it would cover such matters 'as our record of consistent underestimation of the Soviet program, and the Government's lack of willingness to take proper risks,'" the *New York Times* duly informed its readers.

The inquest the administration had feared was now official. "It's a real circus act," John Foster Dulles grumbled. Unfortunately, he added, "the weight of this thing" was deadly serious. Once more Nixon pleaded with Eisenhower to head Johnson off. The vice president had his own electoral future to think about, and he knew that Johnson was no fool. The Texas senator was certain to tarnish him with the same mud-flinging brush he was going to use to paint the entire cabinet as incompetent. But Ike wasn't worried. The American people would see through the populism and demagoguery. "Johnson can keep his head in the stars if he wants," the president replied. "I'm going to keep my feet on the ground."

Nikita Khrushchev, meanwhile, was also doing his bit to stir up trouble for the White House. Sputnik II, he lectured world Communist leaders on the eve of the November 7 celebrations, "demonstrates that the USSR has outstripped the leading capitalist country—the United States—in the field of scientific and technological progress. The launching of the Sputniks undoubtedly also shows," he added triumphantly, "a change in favor of the socialist states in the balance of forces with capitalist states."

Pravda piled on, boasting that "the freed and conscientious labor of the people of the new socialist society makes the most daring dreams of mankind a reality," while the vapid and decadent West wallowed in racial unrest and inequality. America had lost its place in the sun, seemed to be the message from Moscow, and the theme was quickly picked up by American pundits and politicians, who began to worry whether *Pravda* had a point. It was time, warned Senator Bridges, "to be less concerned with the depth of the pile on the new broadloom rug, and to be more prepared to shed blood, sweat and tears if this country and the free world are to survive."

In the pages of the staunchly Republican *New York Herald Tribune*, the financier and statesman Bernard Baruch chastised America for its lack of resolve. "While we devote our industrial and technological might to producing new model automobiles and more gadgets," he wrote, "the Soviet Union is conquering space. If America ever crashes, it will be in a two-tone convertible."

"It's time to stop worrying about tail-fins," Edward Teller, father of the hydrogen bomb, said, continuing the automotive allegory and bemoaning the fact that American culture prized football players above scientists and talk show hosts over university professors. The *New York Times* editorial board agreed, warning, "We've become a little too self-satisfied, complacent, and luxury loving."

America's sense of self, already shaken by the first Sputnik, now foundered in the wake of its much larger, more sophisticated sibling. The Sputniks were "an intercontinental outer-space raspberry to a decade of American pretension that the American way of life was a gilt-edged guarantee of national superiority," suggested Clare Boothe Luce, the millionaire playwright, congresswoman, ambassador, and Republican fund-raiser, whose husband owned both *Time* and *Life* magazines.

"We ourselves have made it an article of faith that the nation which builds the biggest bombs must be morally superior because it is materially superior," she declared. "We need not be surprised today that Russia is making the same claim." And yet, she continued, "we go on believing that our system can provide guns *and* butter. Yes, and Bibles too. But we query whether that means atom bombs *and* bombes glacees; SAC by General LeMay, *and* sack dresses by Christian Dior; lower taxes *and* higher rockets—all this and heaven too."

The Eisenhower administration could not easily brush off the second, more introspective, wave of unease brought on by another Soviet triumph. The first Sputnik had made Americans afraid for their lives; Sputnik II made them question the American way of life. The country was losing faith in itself and in the administration, John Foster Dulles worried. "From the echoes of the satellite have come to me and others from many sections of the country a strong sentiment that the President alone can give the leadership which will restore a feeling of reasonable security and faith in the Administration," one of his aides wrote in a memo that the secretary of state circulated to White House staff. "This leads on every side to the desirability of finding a suitable date, in the not too distant future, to make a strong fighting speech."

Ike liked the idea. There was nothing wrong with America, he believed; it was the greatest, most powerful nation on earth. People had simply caught a case of the jitters and needed a little reassurance. Eisenhower decided to deliver a series of morale-boosting addresses modeled after Franklin Roosevelt's famous Fireside Chats, what White House aides dubbed Operation Confidence, or "Chin Up" speeches, as they became known.

The first of the televised talks occurred on November 7, four days after Sputnik II, and on the same day that newspapers across the country carried another saber-rattling interview with Khrushchev. "The fact that we were able to launch the first Sputnik, and then, a month later, launch a second shows that we can launch ten, even twenty satellites tomorrow," the Soviet leader boasted, neglecting to mention that Korolev had used up his last R-7. "The satellite is the intercontinental ballistic missile with a different warhead. We can change that warhead from a bomb to a scientific instrument," he added, in case anyone missed the point.

In his speech that evening, Eisenhower issued his retort. "The United

States can practically annihilate the war-making capabilities of any other nation," he said, listing the country's lethal arsenal of long-range bombers, fleets of submarines parked under the polar ice cap, and the powerful rockets of its own that were being developed. "We are well ahead of the Soviets in the nuclear field both in quantity and in quality," Ike declared. "We intend to stay ahead."

"Although the Soviets are quite likely ahead in some missile and special areas, and are obviously ahead of us in satellite development," he conceded, "as of today, the overall strength of the Free World is distinctly greater than that of Communist countries."

As the president spoke, the camera panned back, revealing first his Oval Office desk, where a small brass plaque displayed the motto GENTLE IN MANNER, STRONG IN DEED, and then a strange white triangular object on the carpet at his feet. It was a nose cone from one of Wernher von Braun's Jupiter C test rockets, and Eisenhower informed viewers that it had been shot into outer space during successful missile reentry tests. This was evidence, he said, that America was forging ahead with its own space and rocket programs, and that the situation was well in hand. "It misses the whole point to say that we must now increase our expenditures on all kinds of military hardware and defense," the president warned. "Certainly, we should feel a high sense of urgency. But this does not mean that we should mount our charger and try to ride off in all directions at once. We cannot on an unlimited scale have both what we must have and what we would like to have. We can have both a sound defense and the sound economy on which it rests—if we set our priorities and stick to them."

The message was clear: America was safe and strong, and no panicked deluge of defense dollars should be expected from the White House anytime soon. Eisenhower did, however, offer one concession to the new post-Sputnik reality: "I am appointing Dr. James Killian, president of the Massachusetts Institute of Technology, as special assistant to the president for science and technology, a new post," he said, outlining his major initiative to counter the Soviet threat.

• • •

Among the millions of viewers that evening who had expected to hear the announcement of some major American initiative, General Bruce

Medaris watched the address with an equal mix of bewilderment and frustration. Killian—that was it? No new money? No crash programs? No special national priority designations? The general was flabbergasted. Killian, to be sure, was widely respected in the rocket community. As head of the Technology Capabilities Panel that had recommended fast-tracking missiles in 1954, he had been the driving force behind America's belated efforts to compete with the Soviets. But the appointment of a lone academic whom the *New York Times* described as "somewhat cherubic" and "as disarmingly pleasant as a successful hotel manager," hardly evened the score with two Sputniks. Besides, in his new role as a presidential adviser, Killian would not even be in charge of streamlining the various lagging missile efforts. That job was given to an oil company executive with no rocket expertise or technical background.

Damn it, Medaris cursed, pacing around the grand piano that dominated the living room of his ranch house on Squirrel Hill, an official Redstone Arsenal residence that overlooked the Officers' Club, where he had taken Secretary McElroy on "Sputnik Night," as October 4 was now known at ABMA.

Since then no order had come from the new defense secretary authorizing a satellite launch, and Medaris's nerves were shot. He prayed nightly and slept fitfully, driving his Jaguar at breakneck speeds during the day to relieve the tension. But nothing helped. By early November, he could no longer hide his rogue operation from his immediate superior, Army R&D boss James Gavin. "Hang on tight, and I will support you," Gavin had sympathetically urged. "I'm doing the best I can to get a decision."

Gavin, like Medaris, had had enough of Pentagon politics. In fact, he was seriously contemplating quitting the army, following in the footsteps of General Trevor Gardner, the Pentagon's chief missile overseer, who had resigned that summer to protest Donald Quarles's budget cuts. Gavin already had one foot out the door and didn't have much to lose. He was only too happy to rattle some cages on his way out.

At least Medaris now had an accomplice, a coconspirator to watch his back. That gave him some hope. With the humiliation of another Sputnik, he had reasoned, there was no way the Eisenhower administration could continue sitting on the sidelines. Surely ABMA would get its shot now. But not only had the president made no such announcement during his "Chin Up" speech, he had not even mentioned that it was ABMA that had fired the nose cone he had paraded before viewers. "So far as the public could judge, a faceless and nameless group" had done it, Medaris fumed, complaining of the "bitter experience of total anonymity," a state all too familiar to Sergei Korolev.

Medaris was not the only one at ABMA battling frustration. Von Braun was also complaining loudly, only he was doing it publicly, which was not helping the army's case. Gavin's boss, General Lemnitzer, made this clear in a telephone call to Medaris. "The time for talking has stopped," he ordered. Von Braun's outbursts were "causing concern in high places."

The Disney star, in fact, was venting his opinions with such vitriol that the Pentagon had to intercede with the head of the Associated Press to censor some of his more biting remarks, trading the promise of some future scoop to have the comments killed. Sputnik, von Braun had railed, was "a tragic failure for the U.S." Six good years had been irretrievably "lost," he said, wasted while the Soviets had forged ahead with their missile programs. "The real tragedy of Sputnik's victory is that this present situation was clearly foreseeable," he lamented. Saddest of all, he added, was that America had apparently not learned its lesson, since it still wasn't taking satellites or space seriously. "Our own work has been supported on a shoestring while the Soviet Union has emerged more powerful than ever before."

Such inflammatory statements, Lemnitzer warned, "could be very damaging to what the President was trying to do." Ike, after all, was publicly saying that there was no race with Russia and that "no competition in the space field" existed. It didn't look good, then, if the country's best-known rocket scientist went around shouting that there was such a contest, and that the United States was getting trounced. Shut von Braun up, Medaris was told, and the army will take care of the satellite mission.

Von Braun, though, was not alone in contradicting the official line coming out of the White House. As if the president didn't have enough troubles, a national security panel that he himself had convened chose this inauspicious moment to deliver a devastating report that contravened almost everything he said during the November 7 address. Chaired by H. Rowan Gaither, the head of the Ford Foundation, the panel had conducted an exhaustive study for the National Security Council on the nation's state of defense readiness. The upshot of its findings, which landed on Eisenhower's desk only a few hours before he was to go on national television, was hardly optimistic. It recommended the urgent appropriation of an additional $40 billion—an amount equal to the entire military budget—to shore up America's woefully inadequate defenses against possible Soviet missile attack.

The Gaither report detonated like a psychological bomb in the Oval Office. Sherman Adams worried Americans would find it "deeply shocking," and counseled against releasing the study. John Foster Dulles was of the same mind, warning that making the document public would have "catastrophic results." Moscow would perceive it as a sign of weakness, and the Democrats would have a field day undermining the president's position. Already that scoundrel Lyndon Johnson had gotten wind of the report and wanted a copy for his upcoming congressional hearings.

Eisenhower adamantly refused, citing a little-used constitutional clause known as executive privilege. "Its disclosure would be inimical to the nation's security," he flatly told Senator Johnson, who for once found his vaunted powers of persuasion ineffective.

"It will be interesting to find out how long it can be kept secret," Ike later observed at an NSC meeting during which Vice President Nixon argued against burying the study. It would leak anyway, he predicted, and the rumors and excerpts would be taken out of context by the media and would probably sound more frightening than the actual report.

Nixon was right. Snippets from the text began appearing in the press within days, though Chalmers Roberts of the *Washington Post* would break the most complete and alarming account of the study a few weeks later. "The still top-secret Gaither Report portrays a United States in the gravest danger in its history," he wrote. "It pictures the nation moving in a frightening course to the status of a second-class power . . . and finds America's long-term prospect one of cataclysmic peril in the face of rocketing Soviet military might. . . . Many of those who worked on the report were appalled and even frightened at what they discovered to be the state of American military posture in comparison with that of the Soviet Union."

Not surprisingly, the Gaither report undid all of Ike's attempts to re-store calm and order with his Operation Confidence pep talks. "Argu-ing the Case for Being Panicky," retorted the headline in *Life* magazine, in what was essentially a slap in the administration's face. Once more, columnists howled that the president didn't seem to appreciate the gravity of the situation. "Another tranquility pill," one pundit scoffed at the November 7 national address. "It was by no means a blood, sweat and toil speech," commented Eric Sevareid more charitably. "It con-tained little suggestion that sacrifices may be ahead, or that [Eisen-hower] personally thinks they are necessary." Editorials bristled with renewed indignation over the apparent complacency in the White House, which some now called the Tomb of the Well-Known Soldier. "Two Sputniks cannot sway Eisenhower," griped the liberal-leaning *New York Post*. "The President's answer in each instance is the same: we can't do very much."

Amid the barrage of criticism, Eisenhower's approval rating plunged, sinking by 22 percentage points in barely six weeks, an unheard-of pace of decline in the modern presidency, where such erosions usually occur over far longer periods. "In a matter of a few months," noted the histo-rian Walter McDougall, "the rhetoric, the symbology of American poli-tics had left Eisenhower completely behind."

• • •

While Eisenhower tried to ignore accusations of falling behind, pres-sure mounted for him to act more forcefully and to spend more freely. Democrats called for the immediate construction of a national network of air raid shelters, as recommended by the Gaither report, at the horri-fyingly high cost of $22.5 billion. Congressmen demanded an emer-gency infusion of $3 billion to jump-start America's lagging missile efforts, while educational groups exploited the Sputnik panic to push bills that would revamp curriculums, with a focus on science and math-ematics. A measure to inject $1 billion of federal funds into high schools was proposed. "The bill's best bet," one lobbyist slyly noted, "is that the Russians will shoot off something else."

Massive university scholarship programs were proposed to bridge the alleged education gap with communism, and there were even sug-gestions that the federal government begin granting college loans to aspiring students. "Eisenhower was skeptical about the loans," Killian recalled. "He doubted whether young people and their parents would be willing to go into debt for their education."

The White House was also being bombarded with frantic calls for a complete overhaul of the Defense Department and military space organizations, leading to a dizzying array of acronyms competing for presidential attention. A new $100-million-a-year Astronautical Research and Development Agency, or ARDA, was proposed by the American Rocket Society, to coordinate scientific space projects. A rival plan by the Rocket and Satellite Research Panel was even more ambitious, calling for one billion dollars a year to be appropriated for a civilian body called the National Space Establishment (NSE), along the lines of NACA, the existing National Advisory Committee on Aeronautics. Not to be outdone, NACA created its own Special Committee on Space Technology, or SCST, to study the possibility of starting an even costlier space-oriented department, the National Aeronautics and Space Administration, or NASA. Meanwhile, the Pentagon unveiled plans for a new Advanced Research Projects Agency, or ARPA, to conduct military space research and development programs.

All this was exactly what Eisenhower had feared, the panicked spending that would throw fiscal discipline completely out of whack. To the beleaguered president, it must have seemed as if everyone suddenly had a panacea to counter the Soviet space lead, and unfortunately every one of the miracle cures landing on his desk held the unappealing promise of being ruinously expensive. "Look," Ike finally snapped, "I'd like to know what's on the other side of the moon, too, but I won't pay to find out this year."

By this year, he meant the upcoming fiscal year. Treasury Secretary George Humphrey had prepared an austerity budget to weather an economic downturn that was worsening by the month, and he had warned the president that any increases in the already mounting deficit could spark "a depression that will curl your hair." Though Humphrey had recently been replaced by Robert Anderson, the recessionary warnings coming out of the Treasury and Commerce departments had only grown more ominous. Unemployment was expected to jump by as much as 1.5 million in 1958, bringing the total number of out-of-work Americans to 5 million, the highest jobless count since 1940, and economic growth in the third quarter of 1957—after more than a decade of robust increases—had completely flatlined. This was hardly the time to be writing big checks, and Ike was not about to risk a lengthy recession simply to satisfy short-term space jitters.

He did, however, relent on pressures to speed the satellite programs, authorizing the minimal disbursement of $3.5 million (much of which Medaris had already spent) for ABMA to ready its Jupiter C for a space shot. Far from being overjoyed, Medaris hit the roof when he saw the official Pentagon directive. It did not instruct ABMA to proceed with a launch, only to "prepare" for one. "In effect there was no clear-cut authority to go ahead and put up a satellite," he recalled. Vanguard still had the primary mission, and only if it failed in its scheduled December attempt would the White House then consider allowing ABMA to proceed. This was unacceptable, Medaris complained. "They are trying to delude Congress and the public into believing that we are cranking up for a launch."

"Either give me a clear-cut order to launch or I quit," the rebellious general cabled Gavin, declaring rather crudely that ABMA would not carry out its orders. "I'm afraid my language was pretty rough," he recalled years later. But, like Gavin, he was at his wit's end—fed up with the army, the politics, the ceaseless wrangling. He'd walked away from soldiering several times before in his eclectic career, and he had always returned to the military's bosom because civilian life was just too dull. But he was tired of the nonsense, he was frustrated by the low pay that (coupled with his own extravagant tastes) perennially left him in debt, and he was losing his will to fight. More and more he'd been turning to prayers for guidance—Medaris had found religion during World War II—and taken to wondering if his was the right calling. (The religious rebirth, apparently, had not made him any less argumentative; his hard-headed outbursts had simply assumed "a fierce religious zeal" and a "pious belligerence," according to Killian, who was no fan.)

Medaris was not the only one ready to throw in the towel. Von Braun also tendered his resignation. Already his brother Magnus had quit the army to work for Chrysler for twice the pay, and now von Braun was threatening to join him unless ABMA got a green light. Stuhlinger and several other top engineers added their names to the growing list of walking papers, and in Washington a minor panic ensued. The entire army missile program was about to resign, and something had to be done. ABMA, Quarles now decided, would have its shot. But only in January, and regardless of whether Vanguard was successful. For von Braun, that was enough. "Vanguard will never make it," he declared with a certainty that bordered on arrogance.

• • •

By mid-November, it was no longer a secret that the navy's quasi-civilian bid to orbit a satellite was in serious trouble.

Despite the Pentagon's determined assurances that "all test firings of Vanguard have met with success," America's answer to Sputnik was looking decidedly shaky. Vanguard's launch vehicle—a modified three-stage Viking research rocket under the experimental designation Test Vehicle 2—had failed to lift off on five consecutive occasions between August and October because of mechanical malfunctions. Just about everything that could have cracked, leaked, broken, delaminated, depressurized, detached, smoked, sparked, or shorted out had done so with such maddening regularity that Dan Mazur, a frustrated Vanguard project manager and launch supervisor, begged the navy to stop sending him "garbage" instead of rockets.

Like a great many high-end concepts that looked good on paper, Vanguard was proving difficult to put into practice. When the project had first been pitched in 1955, it had won over the Pentagon's satellite selection committee with its imaginative and elegant design and cutting-edge components such as an "almost developed" gimbaled General Electric main engine that swiveled on its own axis. This was a revolutionary departure from traditional steering methods, which employed fins or small side thrusters, and was only one of the many innovations the pencil-thin rocket promised to unveil. "For all practical purposes the Vanguard vehicle was new, new from stem to stern," said Jim Bridger, a navy engineer. "More to the point, it was an awful high-state-of-the-art vehicle, especially the second stage rocket. In the nature of things, the business of developing the vehicle and getting the bugs out so it would work was fraught with difficulties."

No one on the Stewart Committee, the panel Donald Quarles had convened to rule on the different satellite proposals, apparently foresaw the potential pitfalls of picking a blueprint on such a tight schedule. "It was either forgotten, or not understood, that the last ten percent of 'almost developed' missile hardware was the most difficult," recalled Kurt Stehling, Vanguard's propulsion chief. The fact that some members of the Stewart Committee simultaneously drew paychecks from the aerospace companies bidding on the Vanguard contract had also apparently been overlooked. Committee chairman Homer Joe Stewart, for instance, was a paid consultant for the Aerojet-General Corporation, which hoped to manufacture Vanguard's second stage. Senior panel member Richard Porter worked for General Electric, which was building the "almost ready" main engine, while Secretary Quarles's former company, Bell Labs, was a major subcontractor for the rocket's upper stages.

The army's more conservative proposal, on the other hand, had no potentially lucrative new contracts to offer. It relied almost entirely on existing, "off-the-shelf" technology, and essentially swapped a satellite for a warhead on an upgraded Redstone missile. In opting to go with an ambitious, unproven design, the committee also accepted at face value assertions from the Glenn L. Martin Company, Vanguard's Baltimore-based general contractor, that there would be no cost overruns on the project's overly optimistic $20 million price tag.

The problems started almost immediately. Martin had built Viking rockets for the navy since 1948, but because the Vanguard launch vehicle was entirely redesigned, a Viking in name only, everything effectively had to be done from scratch. Within two months of being awarded the contract, Martin, GE, and the other subcontractors began to complain and Vanguard's budget was revised to $28.8 million. A month later, in October 1956, it was bumped up to $63 million. By March 1957, it had risen yet again, to $88 million, prompting the White House to debate canceling the project altogether. "I question very much whether it would have been authorized if the actual cost had been known," Percival Brundage, the director of the Bureau of Budget, wrote to Eisenhower in April 1957. But "abandoning the project at mid-stage," he added, would lead to the "unfortunate conclusion that the richest nation in the world could not afford to complete this scientific undertaking."

Vanguard narrowly won a stay of execution, though the following month its cost further increased to $96 million. Fed up, Charlie Wilson pulled the plug on Pentagon financing, and another scramble for funding ensued, with the CIA and the National Science Foundation chipping in to bridge the gap. By the time Sputnik went aloft (for an estimated $50 million), Vanguard's total taxpayer bill had crossed the $110 million threshold, and the meter was still running.

Throughout the spiraling cost overruns, the schedules for component parts slipped, and assembled test vehicles—the prototypes—were often delivered late and in such shoddy condition that Dan Mazur, the Vanguard project manager, demanded on one occasion that the entire rocket be sent back to Martin "piece by rotten piece." There were moisture problems, poorly located pressure indicator lines, unsoldered wire connections, corroded and leaky fittings, and badly fitted plugs. The GE engine had to be returned because of "wholesale system contamination." Dirt and metal filings were found in the fuel lines. Cracks appeared in the propellant tanks. Batteries failed. Rubber wind spoilers attached to the exterior rocket casing fell off. The gyroscopes were off-kilter, the hydraulic oil resource was plugged, and aluminum chips were found in the hydrogen peroxide gas-generating system.

Vanguard's tribulations were not confined to cost and quality-control issues. Relations between the program's government overseers and private contractors had grown so strained during the delays that at one point navy personnel were denied parking at the Martin lot in Baltimore. The acrimony resulted in what at times was an embarrassing lack of communication: "What! You want to put a ball in that rocket?" a Martin official exclaimed upon hearing that the configuration of Vanguard's satellite had been changed from a cylinder to a sphere that required a new cradle. "Why the hell didn't someone tell us this?"

When the International Geophysical Year opened in the summer of 1957, Vanguard was nowhere near ready. "We're never going to make it in time," Milt Rosen, the program's intense second-in-command, despondently told his boss, John Hagen, the project's overall coordinator. "Never mind," said Hagen, a gentle, pipe-smoking astronomer who was famous for never once losing his temper during Vanguard's developmental ordeals. "We are not in a race with the Russians."

Hagen, however, felt sufficiently pressured to keep up with the IGY timetable that in August 1957 he agreed to try to launch a partial Vanguard prototype, Test Vehicle 2, even though it was "an unaccepted, incompletely developed vehicle." His decision, he wrote in a stern memo to Martin executives, "violated sound principles of operation." But, he conceded, "this is the only way to have at least some chance of maintaining the firing schedule."

TV2 lived up to expectations, failing to lift off five times in a row. It finally got airborne on October 23, but with only one of its three stages operating. TV3, the final Vanguard prototype, was to be the first attempt at firing the entire system, including the troubled GE main engine, the still-experimental second-stage booster, and the third small cluster of rockets bearing the satellite. A far more daunting challenge than its predecessor, with many more moving parts, untested components, and opportunities for malfunction, the TV3 test-firing was supposed to have been conducted in secrecy. Vanguard's formal IGY attempt wasn't slated until midwinter 1958, with TV5, and then only if the TV3 and TV4 tests went off without a hitch. But after Sputnik, White House press secretary Jim Hagerty had prematurely pressed Vanguard into early service by publicly announcing TV3's December launch date. The newspapers had pounced on the announcement, failing to make the technical distinction between a preliminary satellite test and an actual satellite launch, and TV3 was quickly billed as America's official response to Sputnik.

The usually unflappable Hagen had reportedly cringed at the prospect of debuting an unfinished product in front of a worldwide audience. But the damage was done. The administration, whether out of panic or confusion, had placed the hopes of the entire free world on a booster that didn't even inspire the confidence of its own designers.

• • •

Vanguard's travails had not gone unnoticed by Lyndon Johnson's investigators or by the press, which was in a wrathful mood after eating so much Soviet crow and looking for someone to blame. The navy's wobbly satellite bid made a convenient target. "An astonishing piece of stupidity," groused *Time*, disparaging the Pentagon's decision to go with Vanguard. The syndicated columnist Drew Pearson intimated that the Stewart Committee had been "prejudiced" by conflicts of interest.

Hadn't Vanguard been hamstrung by the administration's "penny-pinching," the *New York Herald Tribune* asked Trevor Gardner—lobbing him a loaded question, since he had resigned his post as assistant secretary of defense in frustration over missile and satellite bottlenecks. Gardner agreed that "the funds estimated by Secretary Quarles were totally inadequate," as was the "low-priority status" the program was accorded by the White House. "It is predictable that the project would consistently slip, and this was pointed out to the responsible Administration officials," he added, "including Secretaries Quarles and Wilson."

Engine Charlie Wilson, for his part, was waylaid by the television journalist Mike Wallace. "You clearly underestimated the importance of basic research. Why?" Wallace demanded.

"This satellite business wasn't a military matter," Wilson replied evasively. "It was in the hands of the scientists." Besides, he airily continued, people are panicking over nothing. "They're so cracked loose on Buck Rogers that they're seeing space ships and flying saucers."

"But Sputnik I and II exist!" an astonished Wallace fired back. "They are not flying saucers."

The upshot of the assault on Vanguard was clear: the administration, and ultimately the president, would bear responsibility if the mission failed. Ike was being set up by the press as the fall guy. His poll numbers plummeting, his supporters growing increasingly impatient, the once immensely popular leader was no longer immune to personal attack. Journalists, hunting in a pack, had turned on him. "Implicit in all the criticism was that he was too old, too tired, too sick to run the country," noted Eisenhower's biographer Stephen Ambrose.

Unfortunately for the president, things were only going to get worse. The Preparedness hearings were set to start in less than a week, on November 25, and Lyndon Johnson seemed to have unearthed every single malcontent who ever graced any U.S. missile program. Rumors filtered back to the White House that Johnson was working around the clock, like a man possessed, to get ready for the inquest. To save time, he'd hired a crack team of Wall Street lawyers led by the trial attorney Ed Wiesl and his junior partner Cyrus Vance, and the two had set up shop at the Mayflower Hotel, where a phalanx of disgruntled officials crowded the lobby.

Johnson had also deputized a battery of congressional researchers for his inquisition. "He never asked the head of my organization whether I was available to do this," recalled Eilene Galloway of the Legislative Reference Service. "He simply preempted me and took me over to his committee to work on this subject, and we were working on it from morning to night."

Pounding up and down Senate stairs two at a time, Johnson raced from meeting to meeting, firing off instructions to out-of-breath staffers, who begged the senator to slow down. He had suffered a heart attack in 1955 only a few months before Ike and seemed to be charging headlong into another coronary. "He was really like a dynamo at that time. He was so energized," Galloway remembered. "Everything had to be done in a hurry."

The reason for the rush was that Johnson had not chosen the date of his inquest idly. "The timing was perfect because it grabbed all the attention and hit the public consciousness pre-holidays," Reedy recalled. Johnson had not wanted the hearings "to get mixed up with Christmas" and had purposefully set them in the run-up to the Vanguard launch for maximum exposure. That had left scant time to prepare, line up the witnesses, prep them, and map out the strategy of attack.

As the calendar wound down, Ike braced wearily for the coming onslaught. "Crisis had become normalcy," he confessed in his memoir, recalling the difficult months during the fall of 1957, the lowest point of his two terms in office.

The pressure on the aging president was taking its toll. To outside observers, it appeared that Eisenhower was losing his vitality. He began to mope and seemed distracted. The British historian Leonard Mosley observed, "His aides who sometimes caught him with a faraway look in his eyes soon learned that what he was thinking about was golf."

But even that sole source of escape was becoming a political liability. In the past, Gallup polls had shown that most American voters did not mind Ike's frequent weekday golf outings. To the contrary, his love of the fairways had reinforced his reassuring image as a cool and collected CEO, never too rattled to get in a few holes before lunch. But now, as his leadership was being questioned, the public was less forgiving, and Democrats were avidly painting the president as a modern-day Nero who golfed while America burned. Governor G. Mennen Williams of

Michigan had gone so far as to compose an ode to the president's crisis-ignoring pastime.

> Oh little Sputnik, flying high
> With made-in-Moscow beep,
> You tell the world it's a Commie sky
> and Uncle Sam's asleep.
>
> You say on fairway and on rough
> The Kremlin knows it all,
> We hope our golfer knows enough
> To get us on the ball.

On Monday, November 25, 1957, the relentless pressures, personal attacks, and barrage of criticism finally got to Dwight D. Eisenhower. That morning, as Lyndon Johnson convened his dreaded Senate hearings with a vicious assault on his administration's complacency, Ike repaired to his office to sign papers. "As I picked up a pen," he later recalled, "I experienced a strange although not alarming sense of dizziness." The words on the page in front of him suddenly became blurry. Then he dropped the pen and couldn't pick it up. "I decided to get to my feet, and at once I found that I had to catch hold of my chair for stability." Unable to stand, Ike called his secretary. "Then came another puzzling experience," he continued. "I could not express what I wanted to say. Words, but not the ones I wanted, came to my tongue. It was impossible for me to express any coherent thought whatsoever."

Eisenhower was rushed into bed, and a team of doctors was summoned. "The President has had a stroke," Sherman Adams tersely told Vice President Nixon, summoning him to the White House. "This is a terribly, terribly difficult thing to handle," he said, his tone suddenly more obsequious, once Nixon had arrived. "You may be President in twenty-four hours."

The following day, however, the chief of staff's frosty demeanor had returned, as the doctors had diagnosed the stroke as mild. The tough old soldier, it seemed, had demonstrated that he still had some fight left in him. Though his memory was still fuzzy, and his vision blurred, Ike stubbornly refused to go to the hospital and insisted on working from his bed. By Thursday, he had pronounced himself well

enough to publicly attend Thanksgiving Day services and begin re-
suming his full duties. Adams once more resumed his treatment of
Nixon as a pesky intruder—the surest sign that a semblance of nor-
mality was returning to the White House

• • •

As December dawned, and Eisenhower recuperated from what his
spokesmen insisted was only "a minor brain spasm," the country began
counting the days to Vanguard's long-awaited launch on Wednesday,
December 4. With the promise of the president on the mend and a
chance to even the score with the Soviets, hope once more entered the
national discourse. For the first time since Sputnik Night, America was
upbeat, almost giddy with anticipation. Newspapers prepared special
Vanguard editions, restaurants served Vanguard burgers, and schools
introduced children to the rudiments of rocketry. Sensing the rising
tide of enthusiasm, Lyndon Johnson shrewdly recessed his hearings to
give lawmakers and the public a chance to focus on Florida, where reporters were already filing anticipatory dispatches from the launch facilities at Cape Canaveral.

"The Vanguard tower was clear against a starry sky, two bright lights
glaring at its base and a red beacon shining at its top," the *New York
Times* primed its readers on December 1, three days ahead of the scheduled launch. "From the beach, the Vanguard crane is one of a community of launching structures, some taller, some broader than others. But
the Vanguard clearly has the next billing at the sprawling missile theater
here," whose audience, the newspaper noted, included correspondents
"from as far away as Europe."

On the morning of the fourth, a cool and windy Wednesday, it
seemed as if every major media organization in America had descended
on the large sand dune just outside the test firing range. Dubbed "Bird
Watch Hill," the windswept promontory crawled with television crews
and sound trucks. Scaffolding and newsreel platforms sprouted from
the sand, while radio reporters raced around, shielding their micro-
phones from the steady breeze coming off the Atlantic. From the
dune's trampled crest, an unruly battalion of six hundred photographers trained their long-range telephoto lenses on launchpad 18A,
where a slender, silvery rocket reflected the morning sun.

The moment everyone had been waiting for had come. All along Route A1A, traffic choked the soft shoulders: station wagons with wood-paneled doors, two-ton convertibles, Buicks with big tail fins. The low, square, air-conditioned motels that had been hastily built to cash in on the space craze—places with names like the Starlight, the Sea Missile, and Vanguard Inn—teemed with excited customers.

All over South Florida, eyes were fixed expectantly on the skies over the Long Range Proving Ground at Patrick Air Force Base, as Cape Canaveral was formally known. The rest of the country watched from living rooms, bars, and sidewalks outside stores selling RCA's new color televisions. At network studios in New York, Walter Cronkite and his fellow broadcasters filled the airwaves with all manner of facts pertaining to the Vanguard satellite, the Viking launch vehicle that would carry it, and the Patrick Proving Ground from which it would lift off. Four species of poisonous snakes, viewers were informed, inhabited the 15,000-acre facility. The surrounding mangrove swamps and scrub palmetto forests were home to the nearly extinct dusky seaside sparrow. The base itself, a former naval station, had been turned over to the air force in 1949 and was ideally suited to launch satellites because, at twenty-eight degrees north of the equator, its location offered the easiest shot into space.

Though missiles had been tested at the complex since the summer of 1950, December 4, 1957, was Cape Canaveral's public unveiling, the first time most people had ever seen or heard of America's gateway to the stars. What Cronkite and America did not see, however, was the condition of the Vanguard launch vehicle, which was out of range of network cameras. "The rocket looked unkempt, as if it had been hurried out of bed," recalled the propulsion engineer Kurt Stehling. "It was only partly painted, frost covered its middle, and strips of black rubber wind spoilers dangled dispiritedly from its upper half."

Thirty-mile-an-hour gusts lashed the rocket as the morning passed, a weather front moved in, and chilly journalists grew impatient. By late afternoon there was still no movement on the launchpad other than the howling wind. The reporters stomped their feet to ward away the cold and speculated as to the delay. A valve on the main booster's liquid oxygen feed line had frozen shut, but this the press did not know. By dusk, impatience had given way to frustration and wagering on whether the mission would be scrubbed. Something had to be wrong. This was taking too long. Finally, at 10:30 PM, word reached reporters that the countdown had been aborted and would resume on Friday, December 6. The official reason for the postponement was wind. Cynics on Bird Watch Hill thought otherwise.

. . .

Though the American people were deprived of a launch on December 4, the fledgling ABC television network treated its viewers to another space spectacle that evening. Capitalizing on Vanguard mania, Walt Disney had scheduled the most ambitious and expensive installment of his "Man in Space" series to coincide with the launch date. As usual, Wernher von Braun hosted part of the show, titled "Mars and Beyond." With Ernst Stuhlinger at his side, a slide rule in hand, his bright blue eyes flashing with almost hypnotic conviction, von Braun demonstrated how a spacecraft could reach the Red Planet. It couldn't use conventional propellants, they informed viewers, because of the enormous amount of fuel required for the thirteen-month trip. "A small atomic reactor," Stuhlinger said, pointing to the teardrop-shaped tip of a strange-looking model vehicle, "would turn silica oil into steam and drive turbines." Like von Braun, Stuhlinger affected the efficiently no-nonsense appearance of what the Disney wardrobe department must have envisioned as the engineering look: pale blue dress shirt tucked into conservative gray slacks, no jacket, restrained tie. Slender and balding, with slightly pinched features, he seemed a suitably stern foil to the telegenically boyish von Braun, whose full head of hair, broad shoulders, and penetrating gaze were more befitting of a matinee idol than a mad scientist.

Mars must have seemed a long way off to those Americans who had sat by their radios and TV sets all day waiting for Vanguard to lift off. Yet von Braun sold the "electromagnetically-driven atomic spaceship" as if its flight was not only possible but inevitable. There was something mesmerizing about his infectious enthusiasm, his spare, purposeful movements, the scholarly self-confidence, even the paisley necktie. He had star quality.

If von Braun was at ease in the new medium of television, it was perhaps because his experience in narrating rocketry films dated back more than a decade. Disney viewers didn't know about his wartime experiences, but the U.S. government had been well aware of how von Braun had gained some of his familiarity before the camera.

His audience, back in July 1943, had been considerably smaller, consisting of just four men: Third Reich armaments minister Albert Speer, Werhmacht missile chief Walter Dornberger, Field Marshal Wilhelm Keitel, and Adolf Hitler. The screening had taken place at Hitler's East Prussian bunker, Wolfsschanze, in the very same concrete-lined conference room where a year later the one-armed Count Claus von Stauffenberg would detonate a briefcase bomb in a failed attempt on the Führer's life. Von Braun and Dornberger had come to the Wolf's Lair to tout the V-2's advances and to ensure that Hitler gave the missile a top-priority classification, which would guarantee the timely delivery of scarce supplies like sheet metal. Anticipating the Führer's fondness for theatrics, von Braun had prepared visual aids: cutaway models and film that had been shot using several cameras simultaneously to capture the V-2's flight from every dramatic vantage point. A great deal of effort had gone into the production, which had been filmed by a professional crew, using the newest color negatives. But then there was a great deal at stake. Hitler was far from sold on the V-2, which he had never seen in action. As an old artillery man, he tended to think of the missile as a giant cannon shell and clearly didn't understand the new technology. And he had had one of his infamous "prophetic" dreams, in which the rocket had failed. That, more than anything, had soured him on V-2 production. But von Braun and Dornberger had several things going for them. The Luftwaffe was losing the air war, and it had become clear that Field Marshal Hermann Göring's Junkers would not be able to bomb Britain into submission. If Hitler approved the V-2 as an alternative, it would open the floodgates for hundreds of millions of Reichsmarks in funding and would keep the other weapons programs under Speer's supervision from snatching up valuable component parts. But first, the Führer needed to be sold on the merits of the missile.

After several anxious hours of waiting, the Führer arrived for the appointed interview, and the projector was finally set up. Von Braun and Dornberger quickly stubbed out their cigarettes—Hitler abhorred smoking—and the film started rolling. Von Braun provided running commentary, illustrating improvements on the V-2's thrust, guidance mechanisms, and accuracy. This was his fourth encounter with the German leader, whom he had first met as a twenty-two-year-old engineer in 1934, shortly after Hitler's National Socialists had seized power. At the time, von Braun and Dornberger, then a young army captain, had just begun the Wehrmacht's embryonic missile program. It offered an intriguing loophole around the onerous restrictions of the Treaty of Versailles, which had barred any German buildup of conventional weapons. Missiles, as a completely new technology, had been not included in the list of banned armaments, and the German army had surreptitiously begun scouring amateur rocketry clubs for recruits. Few rocket enthusiasts had wanted to submit to military rigors, but even as a teenager von Braun had realized that only government agencies had the kind of financial resources to make his rocket dreams come true. He had signed on with Dornberger in 1932, the year before Hitler's ascendancy, while he was still a university student. In 1934, Hitler "seemed a pretty dowdy type," von Braun later recalled. But as the country's new leader, he held the national purse strings. "Our main concern," von Braun elaborated, "was how to get the most out of the Golden Calf."

A decade later, the Führer was still not convinced that rockets were Nazi Germany's salvation, despite the vast sums that had been spent on research and development. For von Braun, the July 1943 meeting was especially critical. The war was beginning to take its economic toll on the Third Reich. Money and materials were becoming scarce. Slave laborers were in short supply. Cuts would have to be made. And so with the stakes so high, he and Dornberger had carefully scripted their cinematographic pitch. Liftoff and flight footage had been seamlessly spliced so that the Führer could more easily envision formations of unstoppable rockets hurtling at five times the speed of sound toward defenseless enemy targets. The Führer's questions had been anticipated and answers prepared in advance. Von Braun had dressed carefully for the occasion, making sure to wear his Nazi Party pin on the lapel of one of the somber suits he favored whenever meeting high officials. He had joined the party in 1937, not out of any great conviction for the Nazi cause but because it seemed like the right move for an ambitious young man dependent on state funding. Expediency, rather than ideology, had also apparently figured in his somewhat more reluctant decision in 1940 to accept Heinrich Himmler's invitation of induction into the SS, the Nazi Party's most fanatical killer corps. Von Braun had contemplated declining the offer. But he had been promoting rockets ever since he was sixteen—in department store booths, amateur newsletters, with the military—and he had long learned that every good pitchman covered all the angles, took every advantage that presented itself. SS membership would stand him in good stead with Himmler, the second most powerful man in Germany.

Von Braun had also learned, like any seasoned salesman, when to back off and stay silent. The carefully edited images of the V-2 spoke for themselves. Hitler watched the launch footage, visibly fascinated. He had entered the conference room looking shockingly "unhealthy and hunched-over," Dornberger later recalled, but at the earth-shaking sight of flames roaring from the charging missiles, the Führer suddenly grew animated, more like his old, firebrand self. At one point, after an especially stirring shot, he leaped from his seat, gesticulating wildly, and demanded that immense ten-ton warheads be immediately placed on each V-2. "A strange, fanatical light flared up in Hitler's eyes," Dornberger recalled, terrified to have to explain that the V-2 could carry only a single ton of high explosives. "But what I want is annihilation, annihilating effect!" the Führer screamed.

Hitler was sold. Von Braun got his funding and a top-priority classification. Enough money would eventually be pumped into V-2 production to build the equivalent of sixteen thousand fighter planes that might have changed the course of the war or at least prolonged it. The development of a Nazi atomic bomb would also be curtailed to finance the manufacture of the costly V-2s, which, in the end, claimed a few thousand casualties and had minimal impact on Allied resistance. As Winston Churchill would note, von Braun had helped persuade Hitler to bet the Reich on the wrong weapon. But in 1943, he was awarded the Knight's Cross and the honorary title of professor, Nazi Germany's highest academic distinction. "The Führer was amazed at von Braun's youth," the historian Michael Neufeld observed, "and so impressed by his talent that he made a point of signing the document himself." Himmler, intuitively attuned to Hitler's wild flights of fancy, conferred the rank of SS major on the wunderkind whose rockets would surely win the war. "Von Braun," Neufeld concluded, "essentially made a pact with the Devil."

• • •

As fate would have it, the first American to debrief von Braun after the war was Richard Porter, the GE executive and future member of the Stewart Committee, who was then on loan to Army Ordnance due to his technical background. At that meeting in Germany, von Braun presented a twenty-page memorandum spelling out his potential value to the U.S. military and apparently left a lasting negative impression on Porter. Something about the way von Braun had seamlessly staged his own defection before the fighting had even ended, or perhaps the images of the corpses and skeletal slave laborers found at the young rocket chief's subterranean V-2 factory, must have rubbed Porter the wrong way. (Von Braun's biographer Erik Bergaust makes the point that Porter was instrumental in scuttling the army's satellite bid in favor of Vanguard.) Homer Joe Stewart himself would later confess that some members of his commission might have been "prejudiced," as the media alleged. But not, he would add, for commercial reasons. There could have simply been an unspoken sentiment that American scientists, rather than a group of ex-Nazis, should lead the country into the dawn of a new era.

• • •

Dawn broke calmly over Cape Canaveral on Friday, December 6, as reporters once more staked out their perches on Bird Watch Hill. They did not have to wait long this time. At 10:30 AM, the towering red and white gantry crane slowly pulled away from the launchpad, and a half hour later the sonorous blast of warning sirens filled the air. "T minus five minutes," a distended voice echoed over the loudspeakers positioned throughout the proving ground.

The moment everyone had been waiting for at last was at hand. After two months of intolerable doubt, humiliation, and unaccustomed anxiety, America would finally salvage its pride and show those commies a thing or two about good old-fashioned Yankee ingenuity. Everything would be right after that; the United States of America would be back on track. "One minute," the loudspeaker sounded. The whole country leaned forward, on the edge of its collective seat.

"Ten, nine, eight . . . ," the final countdown began, and at 11:44:59 AM a hoarse, howling whine slammed the Florida coast. Brilliant white flames shot out from under the rocket. Ice crumpled in jagged sheets from its upper stages as the whole booster shook with earth-jarring force. The roar of the engines increased to a piercing shriek, and the last of the umbilical cords dropped way. The rocket shuddered and strained against its moorings. It was moving! It was up, only a few feet, but it was gaining strength. And then Vanguard quivered, burst into flame, and languidly crumpled onto the launchpad, setting off a blast wave felt for miles. "Oh God! No! Look out! Duck!" the spectators suddenly screamed. Then, just as suddenly, there was silence.

11

GOLDSTONE HAS THE BIRD

John Foster Dulles could barely contain his anger. "What happened yesterday has made us the laughing stock of the free world," he snarled, his face flushed.

Vice President Nixon and the other members of the National Security Council nodded bitterly. They had assembled on December 7 to discuss the fiasco at Cape Canaveral, to try somehow to put the humiliating debacle in a positive light. But so far no one had come up with any sort of silver lining to spin to the clamoring press.

"Mr. President," Dulles continued, "I sincerely hope that in the future we do not announce the date, hour, and indeed very minute of any satellite launch until we know for certain it is successfully in orbit. All this negative publicity has had a terrible effect on our international standing."

A plainer truth could not have been spoken. Confidence in the United States was plummeting abroad almost as dramatically as Eisenhower's popularity was sinking at home. The slide was most pronounced in Western Europe, where polls conducted in Britain and France prior to Sputnik's launch had shown that only 6 percent of respondents saw the Soviet Union as militarily superior to the United States. Now fully half of those surveyed viewed America as the weaker superpower, which did not bode well for Dulles's plans to persuade

NATO allies to accept intermediate-range rockets on their soil. He had hoped to pitch the ballistic missile deployment at an upcoming NATO conference in Paris, but Vanguard's charred remains were proving a tough image to overcome. Instead of triumphantly circling the globe, America's vaunted satellite lay ignominiously in a Florida swamp, where it had been flung during the explosion, and continued to emit its baleful beep until a frustrated reporter finally snapped, "Why doesn't somebody go out there, find it, and kill it?"

This was hardly a resounding recommendation for nuclear-tipped IRBMs, and judging by the derisive reaction of the European press, Dulles faced a tough sell. "Oh, what a Flopnik," the London *Daily Herald* laughed. "Spaetnik"—Latenik—the German dailies played off the word. Vanguard should have more aptly been called Rearguard, snickered the French. Kaputnik, Splatnik, Stallnik, Sputternik, Dudnik, Puffnik, Oopsnik, Goofnik, and every other conceivable permutation blared from domestic and international headlines aimed straight at the heart of America's wounded pride. "This incident has no bearing on our programs for the development of intermediate range and intercontinental ballistic missiles, which are continuing to make fine progress," the Pentagon immediately rebutted in a press release, stressing that no military hardware had been involved in the Vanguard mission. But the damage had been done. At the United Nations, Soviet delegates were coyly suggesting that the United States qualified for the technical assistance programs the USSR offered to developing countries. In Moscow, a dead-serious Khrushchev was threatening to target any NATO member accepting U.S. missiles with his own rockets, which, he added snidely, actually worked. "The Soviets are playing this for all its worth," Dulles spat.

"I'm all for stopping such unfortunate publicity." Eisenhower sighed. "But I've no idea how."

The president glanced around the sparingly decorated conference room—a few old oil paintings, some ship models, lots of long faces—looking for suggestions. But his cabinet secretaries, the CIA director, the head of the U.S. Information Agency, and the chairman of the Joint Chiefs of Staff all appeared to be hiding behind a veil of cigarette and pipe smoke. Jim Hagerty, the White House press secretary who had prematurely released Vanguard's launch date, was nowhere to be seen.

Ike was on his own, that much was clear. The brief reprieve he had been granted after his stroke was over. Hope had been dashed, Vanguard Fries had been stricken from the nation's menus, replaced by Sputnik Cocktails—one part vodka, two parts sour grapes—and the vengeful media, having angrily crowned Vanguard "our worst humiliation since Custer's last stand," were searching for scapegoats. Already, the Glenn L. Martin Company, Vanguard's general contractor, had been punished. Its stock had taken such a beating that it had been forced to suspend trading. Vanguard's project manager, the affable John Hagen, had been equally assailed at a raucous press conference. "This program has had unprecedented publicity in the development stage, which is not usually the case," he said as he tried to defend himself. "The fact that it was a test phase was lost sight of," he added, deflecting culpability from his scientists to politicians, whom he refused to name. The Democrats had no such reservations, eagerly informing voters exactly where to place the blame. "It lies with the President of the United States," Senator Henry "Scoop" Jackson announced haughtily on NBC's *Meet the Press*. The president's lack of vision, obdurate penny-pinching, and baffling complacency, he groused, were directly responsible for the disaster in Cape Canaveral. "How long, how long oh God, will it take us to catch up with the Soviet Union's two satellites?" wailed Lyndon Johnson, who had hastily reconvened his Preparedness hearings to capitalize on the national fury, and like Jackson he was going to point the finger at the Oval Office.

After only a few days' rest at his beloved Gettysburg farm—just enough time to regain what he described as 95 percent of his speech and motor skills—Ike had been thrust into yet another maelstrom. Now the Democrats were insinuating that Eisenhower no longer had the strength and vitality to lead the nation and that he should step down. "There were open and widespread suggestions that the President resign," *Time* magazine reported in its December 9 issue, noting that NATO leaders were "shaken to the point of dismay" to learn that he might not be able to attend the big Paris summit in mid-December. "It is the whole free world that is sick in bed with Ike, waiting for his recovery," a French newspaper commented.

What very few people realized, outside of the president's most trusted circle of advisers, was that Eisenhower was asking himself the very same thing. "In my mind was the question of my future fitness to meet the rigorous demands of the Presidency," he later confessed. "The test that I now set for myself was that of going through with my plan of proceeding to Paris." The Paris conference was a week away, the first ever meeting of all the NATO heads of state, rather than the customary gathering of defense and foreign ministers, and the largest gathering of Western leaders in Europe since the Paris Peace Conference in 1919. It was hugely important not only because of the proposed deployment of American nuclear missiles on the European continent but also because the United States needed to restore the shattered confidence of its anxious allies. Ike had to attend, not only for himself but for the good of the nation as well. "If I could carry out this program successfully and without noticeable damage to myself," he vowed, "then I could continue my duties. If I felt the results to be less than satisfactory, then I would resign."

Nixon and Dulles had not been privy to the president's private pledge. They were not his confidants, like the ferociously loyal Sherman Adams or General Goodpaster, and while the vice president had yet again earned praise for the way he had handled himself during Eisenhower's incapacitation by not appearing too eager to fill his shoes, his low profile was partly calculated. Nixon was purposefully distancing himself from the president because he had no intention of going down with Ike's sinking ship. If the president dreamed of golf and retiring to Gettysburg, that was fine, even understandable. But Nixon was still only forty-four years old, and he had his own future to think about. He had paid his dues, and he had suffered untold slights and humiliation, all so he could one day sit at Ike's desk. And now these blasted missiles were threatening to drag him down too. Slowly, imperceptibly at first, and then demonstratively, the vice president began inching away from his stricken running mate. (The move would inspire a running joke: Martians land in Washington and approach the vice president. Take us to your leader, the aliens demand. "I can't," Nixon demurs. "I hardly know the man.")

Nixon was not the only administration official wondering whether the leader of the world had lost confidence in himself. "This man is not what he was," Adams confided to James Killian, in a rare moment of doubt. The change was dramatic. Only a day earlier the president had seemed on the road to recovery, determined to prove to the country and to himself that he could lick "this cerebral thing." Everyone in the White House had noticed a new energy, a spring in Ike's step, a fighting spirit reminiscent of his first-term buoyancy. That rekindled vigor was now gone. It was as if Vanguard had sucked all the wind out of his sails. Looking around the room at his ambitious vice president, the powerful Dulles brothers, Secretary McElroy, and the assembled NSC staff, Eisenhower must have indeed seemed a shadow of his former self. Once the supreme commander of the greatest fighting force ever assembled, he was now frail and exhausted, an old man presiding uncertainly over a jittery country. If there was a low point, a single, most downcast occasion in Ike's long career in public service, this was almost certainly it.

After a few seconds of uncomfortable silence, Donald Quarles hesitantly cleared his throat. Since he had appointed the Stewart Committee and had ultimately chosen Vanguard, this was his mess. "I think, Mr. President, that in a sense, we were hoisted by our own petard yesterday," he said, launching into another one of his impassioned justifications and rationalizations. The United States had committed itself to share all data from Vanguard with the IGY, he said, and had included the launch date so that scientists in other countries could have their ground stations ready to receive signals from the orbiting satellite. (What he conveniently failed to mention was that the December 6 test flight had never been intended as an official IGY launch.) "I'm not trying to make excuses for what happened yesterday, Mr. President," he added, "I'm just trying to explain why we are obligated to publicly announce launches."

"Do we have to do this in the future?" interrupted John Foster Dulles, whose "irritation" was clearly recorded in the minutes. "The Soviets kept their launches secret, why couldn't we?"

The unfortunate Quarles, who had recommended that Eisenhower downplay Sputnik just prior to his disastrous October 9 press conference, and then compounded the error by overestimating Vanguard's chances of success, would find little respite from the hot seat in the coming weeks. Again he tried to explain that the United States had formally committed itself to conduct satellite research as a purely peaceful, scientific pursuit, with complete openness. "It would involve fundamentally changing our policy," he stammered.

"Well, maybe we should change our policy," Dulles shot back. "Yesterday was a disaster for the United States."

• • •

While America plunged into a national funk and Vice President Nixon scrambled for high ground, the man responsible for all the turmoil lay exhausted in a sanatorium in southern Russia. Sergei Korolev's fragile health had finally caught up with him. During the marathon preparations for Sputnik II's rushed launch, he had taxed his delicate system to the breaking point and had collapsed shortly after completing his mission. He was rushed to Moscow's most exclusive hospital, one reserved for ranking party officials, where the Soviet Union's leading heart specialists were summoned to his bedside. The Chief Designer was diagnosed with arrhythmia, coupled with "over-fatigue," and the doctors prescribed thirty days of bed rest. Khrushchev himself issued the directive, since Korolev had a penchant for ignoring his physicians' advice.

Despite the order to stay off his feet, Korolev had no intention of remaining idle. Kislovodsk, the spa he chose for his enforced recuperation, had been popularized a century earlier by the poet Aleksandr Pushkin and frequented by the czarist aristocracy for the medicinal powers of its warm sulfuric springs. Nestled between the Caspian and Black seas at the foot of the snowcapped Caucasus, it was the burial site of Friedrich Tsander, Korolev's first real mentor and the man who had opened his eyes to the possibility of space travel in the late 1920s.

With his customary zeal, Korolev had turned the peaceful sanatorium into a bustling base from which to launch a region-wide search for Tsander's grave. Local officials, clerical administrators, archaeologists, and even experts from Moscow were summoned, cajoled, threatened, berated, and rewarded until every cemetery in the area had been scoured and the rocket pioneer's final resting place at last found. There, Korolev ordered a monument erected in homage to the visionary who had planted the seed for the Sputniks.

In Moscow, meanwhile, Valentin Glushko and some of the Chief Designer's other envious detractors were taking advantage of his absence to sow seeds of doubt about the R-7 within the Kremlin and the military. For all of Sputnik II's political and propaganda achievements, the mission had not been a technical success. *Pravda* and its wire-service sibling, TASS, had made much of Sputnik II's living payload, regaling readers both at home and abroad with tales of the mixed-breed terrier Laika, hurtling through space 1,000 miles above the earth's surface at a speed of 17,600 miles per hour. Telemetry readings, the public had been told, had shown that her heart rate had jumped dramatically during liftoff, to 260 beats per minute, but she had settled down, enjoying her gelatinized treats as she paved the way for human interplanetary travel in her climate-controlled capsule. In fact Laika had died shortly after launch, when both the heat shields and the cooling systems had failed, and she had suffered a horrific fate akin to being slow-roasted alive in a convection oven. The tragedy would be kept secret until after the collapse of communism, but within the Soviet scientific community it cast grave doubt on Korolev's plans to send human beings into space.

Similarly, the Soviet press had gone to great lengths to publicize Sputnik II's other significant contribution to scientific exploration, its onboard meters that would map the radiation belts that were believed to surround the earth. To better impress the West with its technological prowess, Moscow released astonishingly detailed descriptions of the devices, and yet after the launch no major announcement of glorious discoveries had been heralded by Moscow. The truth of the matter was that Sputnik II had been such a rush job that half the systems on the satellite had malfunctioned. It had performed its political mission, but not much else.

The military, especially, had been unimpressed. The more Marshal Nedelin grew familiar with the slow-loading R-7, the less he wanted it for his Strategic Rocket Forces. Glushko had long been whispering in his ear that the rival R-16 would be better suited for warfare, and in the Chief Designer's absence Nedelin had finally taken the matter to Khrushchev.

"Tell me, Sergei Pavlovich," the Soviet leader confronted Korolev upon his return to Moscow, "isn't there some way we can put your rocket at a constant readiness, so that it can be fired at a moment's notice in the event of a crisis?"

"No," Korolev conceded. But he reacted furiously to suggestions that his missile used the wrong propellant. Cryogenic oxidizers like liquid oxygen, he told Khrushchev, were much safer than the highly toxic and inherently unstable acid mixes proposed by Yangel and Glushko, which he called "the devil's venom." The Chief Designer stubbornly dug in his heels, and he switched the topic to space, his favored distractive ploy, igniting the first secretary's imagination with tales of the 5,000-pound Sputnik III he wanted to launch in the new year; the lunar probes he would send shortly thereafter; and the manned space missions that would follow. Beguiled by visions of political glory, Khrushchev momentarily forgot the fuel issues and in a burst of enthusiasm begged his star scientist "to deliver," in Korolev's words, "the Soviet Coat of Arms to the Moon."

But Nedelin and the military would not let the matter rest. The word was spread that the Chief Designer had lost his bearing, had become too wrapped up in space, and had lost all sense of priority. "Korolev works for TASS," making newspaper headlines, the Red Army chieftains grumbled, whereas "Yangel works for us." They decided to throw their support behind the rival designer. Nedelin would arrange another audience with Khrushchev, this time with Glushko alone.

• • •

As the Soviet military plotted its revolt against Sergei Korolev, the Army Ballistic Missile Agency was feverishly preparing to leap into space. ABMA had shed the requisite crocodile tears for Vanguard's fiery demise, making the expected public statements of support and sympathy, but privately there had been widespread relief in Huntsville that the competition had failed.

Things were finally looking up at ABMA, whose fortunes seemed inversely related to the White House's woes. In addition to finally getting its satellite shot, ABMA had just received a huge boost from the intense pressure brought by Lyndon Johnson's hearings. Johnson, on November 27, had called McElroy and Quarles to appear before his subcommittee. Expecting a politically charged tongue-lashing, the secretary of defense began his testimony by throwing Johnson a bone in the hope that he might deflect accusations that Charlie Wilson's missile cuts had put America in danger. "Before you begin your questioning, I have a brief statement to make," McElroy said. "We have been undertaking during the past few days an intensive re-assessment of our position," vis-à-vis Wilson's directive to eliminate duplicate Army and Air Force IRBMs. "We are today authorizing the placing into production of both the Jupiter and Thor missiles."

Privately, McElroy was less than thrilled with the concession. "The chief reason" for salvaging the Jupiter, he said, was "to stiffen the confidence and allay the concern of our people."

A similarly preemptive tactic was used to co-opt Medaris's potentially damaging testimony. Johnson obligingly played along with the ploy, which took some of the sting out of his hearings, but just as important allowed him to share in the credit for shoring up America's missile program. The army, a smiling and decidedly more docile Medaris informed the subcommittee, "was being authorized to proceed on a 'top-priority' basis with the development of a solid-fuel missile," the Pershing, a storable, next-generation rocket that could be launched on a moment's notice.

All of Huntsville rejoiced at the twin coups. Not only was Jupiter officially and irrevocably saved but ABMA, at long last, had a new assignment. It wasn't going to be disbanded, its staff wasn't going to be snatched up by the rapacious air force, and von Braun's designers weren't going to be limited to any more idiotic 200-mile-range rules. "With feelings much different from those that had had my head bowed and my spirit beaten a year before," a rejuvenated Medaris now set about pressing his advantage and capitalizing on ABMA's rising prospects.

If von Braun were to launch his satellite without a hitch, ABMA would be ideally positioned to take the lead in America's space effort. The door to the heavens had been flung open, and space was now a legitimate political destination. In Congress, Senators Albert Gore of Tennessee and Clinton Anderson of New Mexico had introduced bills to place all space programs under the Atomic Energy Commission. At the White House, Vice President Nixon was said to be especially receptive to proposals for creating a single national space agency. It was clear that the Russians had one and intended to send a man into orbit, so despite Eisenhower's misgivings the administration needed to plan ahead. Medaris wasn't going to let an opportunity like this slip by. He ordered von Braun, whose sole brush with space to date was the nose cone Eisenhower had displayed during his pep talks, to devise a comprehensive road map for America's future cosmic conquests. Von Braun's $21 billion blueprint, cumbersomely titled "Proposal for a National Integrated Missile and Space Vehicle Development Program," envisioned orbiting an astronaut in 1962 and putting a man on the moon by 1970. The plan hinged, naturally, on ABMA's central participation.

Medaris was equally anxious to explore opportunities in the potentially lucrative new field of spy satellites, which, like space, was now also wide open. Since the air force had shown so little enthusiasm for developing its WS-117L reconnaissance satellite platform, there was an opening to grab the mandate for the army. General Bernard Schriever, Medaris's archrival at the air force's missile command, was too busy trying to get the Thor and Atlas operational, while shuttling "like a yo-yo," in his own words, between congressional appearances and his West Coast offices. He might not even have time for satellites, Medaris reasoned. But after a few discreet inquiries, ABMA's liaison officer at Schriever's Los Angeles headquarters "found the door completely shut." The air force, all of a sudden, had developed a proprietary interest in the WS-117L. "Sputnik woke us up," Schriever later conceded, and he wasn't sharing any information with his rival. Medaris responded in the petty spirit of interservice rivalry: "So I also closed the door and told our people to give the Air Force no information on our satellite plans." It was juvenile and "preposterous," he admitted in retrospect, but he couldn't help himself.

Unbeknownst to Medaris, there was a reason for the newfound secrecy; a third party also coveted the WS-117L: the CIA. Richard Bissell had been eyeing the project ever since he had started searching for a replacement for the U-2. He had helped fund Vanguard from his slush fund and tried to covertly buy the Itek Corporation, an optical research laboratory in Boston, which was working on recoverable cameras that could operate from outer space. In the summer of 1957, Bissell, Edwin Land, and James Killian had begun hatching a scheme with Schriever for the CIA to assume direct control over spy satellites, as it had done with spy planes. Schriever was amenable because his missiles would be used to launch the CIA satellites, and he could still play a significant role in the operation. The idea was brought to General Goodpaster at the White House, who had not thought the timing right to approach the president with the plan. After the launch of Sputnik I, Eisenhower himself broached the topic and asked for a briefing on reconnaissance satellite developments.

The problem with the WS-117L was that it relied on video image transmission, a technology that was still embryonic and would not be perfected for many years. Land and Killian were proposing an interim solution: cameras similar to those used on the U-2 would be launched by a two-stage Thor into orbit, where they would snap shots of Soviet targets and jettison canisters of film. The negatives, in heat-resistant containers, would fall back to earth at predetermined locations and deploy parachutes that could be recovered in midflight. Momentum for the proposal grew in November, as Sputnik II increased the sense of urgency that the WS-117L needed to be fast-tracked, and that the air force's bureaucracy simply moved too slowly for the job. The Vanguard fiasco finally gave the CIA the opening it needed. In the stunned aftermath of the explosion, Allen Dulles, Jim Killian, and Neil McElroy quickly convinced Eisenhower to secretly promote the reconnaissance satellite to a "national security objective of the highest order," a prerequisite for Bissell's friendly CIA takeover.

The plan was set in motion over the next few weeks. "Our first goal was to put the genie back in the bottle," Bissell recalled. The air force photoreconnaissance program had received far too much publicity; the *New York Times* had written about it in front-page stories, and Johnson's subcommittee had discussed it in open sessions. The project would have to be canceled, and with as much fanfare as possible. They would pick a slow news day over the next few months for the Pentagon to make the announcement. Cost overruns, technical difficulties, or some other excuse would be invented. Outside of a few top generals like Schriever, not even the air force would be told the real reason. The program would be restarted on the sly under a new code name. "I had to invent an elaborate cover explanation," Bissell recalled. Finances and procurement would be handled through bogus departments and fictitious front companies in much the same vein as the U-2. "We also had to have a plausible cover story for that part of the project that couldn't be hidden from the public," Bissell said. The frequent launches from Cape Canaveral would be explained by an IGY-inspired civilian research program that would build genuine research satellites and produce reports and studies. They would call it Discover, which had a peaceful, scientific ring. Its real code name would be Corona, after the typewriter on which Bissell outlined the takeover scheme. It would be the most ambitious, secretive, and costly operation in CIA history. If all went well, that is.

. . .

Richard Bissell and Bruce Medaris were not the only ones to find a silver lining in Vanguard's implosion. The catastrophe was also serving Lyndon Johnson well, as it presented fiery evidence of American missile missteps and focused public attention on the Preparedness hearings.

Never one to pass up a media opportunity, Johnson played the disaster for all it was worth. John Hagen, the soft-spoken Vanguard program director, was hauled in as a witness and bludgeoned until he confessed that funding shortages had contributed to his rocket's less than spectacular debut. Johnson then set his sights on military missile programs, which he claimed suffered from the same penury, exposing America to the terrifying might of Soviet rockets. "Some awful needles were stuck into this thing," George Reedy chuckled apologetically decades later. "I can still remember the hearing when we left with a distinct impression that the Soviets outnumbered us by a factor of fifteen to one. We were giving them credit for maybe 1,500 missiles and we were only supposed to have thirty."

Thus a new gap, "a missile gap," was born from the ashes of Vanguard. "We will be walking a very tight wire with our lives for the next five years," a senior executive from General Dynamics testified, explaining how the late start in developing the Atlas ICBM his company was building meant that for the foreseeable future the United States would have to rely on planes to defend itself against the Soviet Union's virtually indestructible new missiles.

The administration, it went without saying, was responsible for the strategic imbalance that now imperiled the nation. For all his bipartisan pledges—his noble talk of there being no Republicans or Democrats after Sputnik, only Americans—Johnson had effectively put Eisenhower's entire government on trial. The indictment was all the more devastating because it was subtle. Johnson studiously avoided histrionics and tended to chide gently, more the reproachful schoolteacher than the vengeful prosecutor. "There are too many people in government who have the right to say no," he admonished. "Too few who have authority to say yes, and even less who dare to do so." Johnson never directly pointed the finger at the White House. That was left to others—supposedly impartial scientists, who decried Eisenhower's "false economies" and wrongheaded policies, or friendly journalists, who railed against the grave dangers the country faced as a result. "At the Pentagon they shudder when they speak of the 'gap,'" reported the columnists Joseph and Stewart Alsop in one of their more alarming dispatches from the hearings. "They shudder because in these years, the American government will flaccidly permit the Kremlin to gain an almost unchallenged superiority in the nuclear striking power that was once our specialty."

Few readers could have missed the implication. Ike was "flaccid," rolling over for the Russians. Johnson, on the other hand, stood strong and tall, seeking truth and security at a time when America, as *Harper's* magazine metaphorically put it, was "a leaky ship, with a committee on the bridge and a crippled captain sending occasional whispers up the speaking tube from his sick bay."

In his pursuit of publicity—and perhaps even in pursuit of truth and security—Johnson was relentless, shuffling the attending representatives of the press like a circus master. Photographers would be ushered into one chamber, shots snapped, then hustled out to make room for the reporters. "Speaking so fast that no one could take a word-by-word account," observed the historian Robert A. Caro, "he would rip through a briefing on a committee session, pant that he was ten minutes late for a luncheon speech he had to make. 'The statements will be up in a minute anyway,' burst out of the room to give the television interviewers time for 'just three' questions, then flaring up when a fourth was asked—'I told you, just three,'" he would add, before running down the hall with a dozen reporters in tow.

Grabbing the next day's newspapers, Johnson would scream, shout, cheer, sob, curse, and vow vengeance, depending on how his heroics had been portrayed. Then he would call a press conference and start all over again. "Control of space means control of the world," he would warn, posing dramatically for the cameras. "From space the masters of infinity would have the power to control the earth's weather, to cause drought and flood, to change the tides and raise the levels of the sea, to divert the Gulf Stream and change the temperature climates to frigid." Nothing short of planetary domination hung in the balance of "this ultimate position," Johnson railed against America's seeming inability to achieve orbit. "Our national goal and the goal of all free men *must* be to hold that position."

"Light a match behind Lyndon and he'll orbit," cynical journalists joked about the senator's harried pace. But the public was not so jaded. America was watching and listening, and Lyndon Baines Johnson was making sense.

• • •

The one nugget of information the Johnson subcommittee could not mine from its witnesses was when the United States would attempt to orbit another satellite. "Soon" was all Medaris would say when pressed in his testimony on January 7, 1958.

"I am not going to ask you about the precise date," Cyrus Vance persisted, seeking to pry a more exact answer out of the general. "I am thankful for that, Sir," Medaris replied, not taking the bait. To the launch crew at Cape Canaveral, he wired the following instructions the next day: "Do not admit to the presence of the vehicle. Shroud upper stages with canvas and move to the pad not later than 6:30 A.M."—that is, under the cover of darkness—"Identify the vehicle as a Redstone. Great care should be taken concerning the movements of key personnel from the agency in your vicinity. They will be flown directly by special plane. Any violation of this decoy plan will be dealt with severely."

To his own staff at ABMA headquarters, Medaris issued similar injunctions against discussing any aspect of the pending launch, even with their wives. "I desire it well understood that the individual who violates instructions will be handled severely," he reiterated. The lessons of Vanguard had been well learned. There would be no advance notice this time. It wasn't just the press that Medaris was worried about. The last thing he needed was a herd of self-aggrandizing politicians descending on Huntsville and the firing range in Florida. Not only would they be a distraction, bringing down hordes of pesky journalists and putting unnecessary pressure on his whole team, but their mere presence could also inadvertently scuttle a launch. "Personal observation had convinced me that the chances of success on any important firing effort were in inverse proportion to the number of VIPs present," Medaris later explained. With half of Washington looking over their shoulders, ABMA's launch crew would be reluctant to scrap or postpone a shoot because of inclement weather or minor technical glitches, which could prove disastrous. Medaris knew that "there was every human tendency to decide in marginal cases to go ahead and accept the risk rather than disappoint the visitors."

Medaris was not going to allow that to happen. The VIPs would remain in Washington and would be kept in the dark like everyone else. Only a few people in the Pentagon and at the National Security Council were told that liftoff was scheduled for Wednesday, January 29. To further mask its activities, ABMA began referring to the satellite booster in all official communications simply by its serial designation: Missile Number 29. Missile 29 was one of the original Jupiter Cs that Medaris had quietly diverted during the 1956 reentry tests "for more spectacular future purposes," as he had hopefully put it. Taken out of cold storage, the test missile had been completely disassembled in late November. It consisted of four stages. The main stage was an elongated Redstone. Eleven scaled-down Sergeant rockets formed the second stage. Three Sergeant motors formed the third stage, while the satellite would be embedded into the final Sergeant rocket in the fourth stage. Von Braun's team had worked day and night throughout December to reconfigure the four-stage carrier. A new fuel, hydyne, had replaced alcohol to increase thrust from 75,000 to 83,000 pounds, and the turbo pumps had been upgraded to work longer, extending the Redstone's burning time from 121 to 155 seconds. The entire forward section, which housed the upper stages, had also been modified to accommodate a special "spinning bucket" that spun on its axis, creating a gyroscope that would keep the cluster of smaller top-stage rockets in perfect equilibrium. An improved inertial guidance system was installed, as were a series of tiny directional air-jet nozzles that would keep the uppermost stages perfectly aligned with the earth at the point of orbit. Everything was tested and retested until von Braun pronounced himself satisfied. "Ship it to Florida," he declared on December 20. "It will do the job."

One final modification was made to the Jupiter C booster after its arrival at Cape Canaveral in the belly of a specially configured C-124 cargo plane: its name was changed to Juno. The rechristening was ordered by the Pentagon to deemphasize Missile Number 29's military and Germanic origins. The Jupiter C's main stage, after all, was a direct descendant of the V-2, and it was said that some folks in Washington wanted the lineage obscured—and a female name would accomplish this task quite nicely. Medaris, however, suspected baser motivations. "It became quite obvious that every effort would be made at the national level to suppress the Army's participation in this enterprise," he worried, "and to credit the whole business to the scientific personnel controlling the IGY effort."

Juno did have civilian components. Its uppermost stages had been designed by the Jet Propulsion Laboratory at the California Institute of Technology, and JPL had also made the eighteen-pound satellite, which was fitted into the crown of a slender six-foot-long Sergeant rocket that would ignite for six and half seconds just prior to orbit. The satellite, in fact, had been secretly built several years ago, when JPL's director, William H. Pickering, had conspired with Medaris to circumvent Washington's decision to go with Vanguard. Without authorization, they had proceeded with work on an army satellite on the sly, "just in case" Vanguard failed. "We bootlegged the whole job," Pickering later admitted. "When we finished we locked up the satellite in a cabinet so it wouldn't be found."

That earlier act of insubordination proved not only prescient but also hugely time-saving, as the contraband satellite was taken out of hiding and put directly into the rocket. Like the first Sputnik, it contained two tiny radio transmitters to relay data to ground stations. A miniaturized Geiger counter to measure cosmic radiation was added by James Van Allen, the renowned astrophysicist from the University of Iowa, who in 1950 had first proposed holding the IGY in 1957. Van Allen had originally designed his radiation metering device for Vanguard, but he "thought it would be wise to prepare it in such a way that it would fit Vanguard as well as Jupiter C so that [he] would be prepared in either case."

Though Juno's civilian contributions were not insignificant, it became abundantly clear to Medaris from the wrangling over the classified press releases that were being prepared ahead of the launch that JPL and the IGY committee would get a disproportionate share of the postorbital credit. "Almost every reference to Army-developed hardware was stricken from these documents," Medaris fumed, "in a rather dishonest attempt to make our first space triumph look like a civilian effort."

A myth was being born: that the conquest of space had been driven by man's insatiable appetite for exploration, rather than by the arms race. Even the satellite's new name, Explorer, bore witness to the elaborate PR campaign quietly being prepared in Washington.

Of course none of this information would ever be released to the public if the launch failed. If the launch failed, it would be the army's fault, and ABMA would go back to making weapons of mass destruction. "This is our biggest challenge," Medaris confided to his wife in a rare moment of doubt. "We've waited a long while for recognition and now we must make good on our promises. . . . I'm praying for help."

• • •

Few public figures in Washington could have used as much help as Dwight Eisenhower in January 1958. His approval ratings had fallen another eight points after the Vanguard debacle, bringing the precipitous slide to a total of thirty percentage points in a little over four months, and the beleaguered former general was still not himself, uncertain if he could carry out his duties. But on a positive note, the trip to Paris had gone surprisingly well. The British and the Italians had agreed in principle to accept Thor intermediate-range missiles as part of NATO's defense shield, though now the United States would also have to find a home for the additional Jupiters it had agreed to produce to satisfy Lyndon Johnson's subcommittee. More important, Ike had impressed his hosts with a combative attitude that belied his weakened medical condition. He was once again the war hero of old that the Europeans remembered.

But somewhere over the Atlantic, the president once more lost his fighting spirit. When he returned to Washington and appeared with John Foster Dulles at a televised press conference to report on the NATO summit, he seemed listless and deflated. Dulles did almost all the talking, while Ike, at times, looked completely detached and uninterested as his secretary of state droned on. After observing the joint television appearance, Harry Truman quipped that he had been "just about as thoroughly bored with Mr. Dulles as the President was." The press pounced with renewed calls for Eisenhower's resignation. *Time* did its bit to further deflate the sinking American leader. "The symbols of 1957 were two pale, clear streaks of light that slashed across the world's night skies and a Vanguard rocket toppling into a roiling mass of flame on a Florida beach," it noted in its year-in-review issue. "On any score 1957 was a year of retreat and disarray for the West. In 1957, under the orbits of a horned sphere and a half-ton tomb for a dead dog, the world's balance of power lurched and swung toward the free world's enemies. Unquestionably, in the deadly give and take of the cold war, the high score of the year belongs to Russia. And, unquestionably, the Man of the Year was Russia's stubby and bald, garrulous and brilliant ruler: Nikita Khrushchev."

As the American press hailed Khrushchev's ascent with grudgingly glowing cover stories, Eisenhower was quietly conferring with his attorney general to "make some specific arrangements" for the vice president to succeed him in the event of further incapacitation. He was also preparing a highly unusual State of the Union address in which he would concede that 1957 had been "no ordinary" year. "I decided to confine the annual message—probably for the first time in history—to just two subjects," he said, "the strength of our nation: particularly its scientific and military strength, and the pursuit of peace."

The State of the Union would be one last "Chin Up" talk, using the biggest stage afforded to an American leader to try to put the nation's plight in perspective. He would stress the country's considerable resources, talents, and relative merits, and he would outline specific plans to shore up what he considered minor education and defense shortcomings. He would rally the troops, as he had done on D-Day and at the Battle of Bulge. And he would try to steal Lyndon Johnson's thunder in doing so.

But America was in no mood to listen to its old soldier. The public did not want to be placated with soothing words. Words were empty. What America wanted was action, the sort of call to arms that the flamboyant senator from Texas was advocating. A low point had been reached where no amount of reassuring would restore the country's shattered confidence, either in itself or in its commander in chief. Only a successful satellite would make things right again. The only question was which satellite: the army's or the navy's?

• • •

Vanguard's TV3 might have died a very public death, but the $110 million program behind it was still very much alive. The administration simply had too much invested in Vanguard, in terms of both financial and political capital, to pull the plug just because of one highly publicized failure. In the new spirit of discretion demanded by John Foster Dulles, launch dates were now classified. But otherwise, the six-vehicle program remained unchanged. It still enjoyed priority over ABMA's Explorer satellite, and on Thursday, January 23, it was given one last shot to beat the army into space.

The rival teams eyed each other warily at the increasingly crowded Cape. From their vantage point at launchpad 18A, the Vanguard crew watched anxiously as von Braun and his rocket team began setting up shop at launch complex 26A, a few hundred yards away. "We could see the Army preparations on their launch pad not too far from us," recalled the Vanguard propulsion engineer Kurt Stehling. Like his competitors, Stehling had been born in Germany. But rather than work for the Nazis, his family had fled to Canada, and he later emigrated to the United States to pursue his space dreams. Now, as he looked over his shoulder at the elongated Redstone being erected nearby, he bitterly reflected how the army's "warhorse" rocket held an unfair advantage over his "skittish thoroughbred" because its "progenitor was built in Germany" at a cost of thousands of lives. But then justice and morality had no place on the launchpad; science was blind that way.

TV3BU, TV3's designated backup vehicle, was proving as skittish as its late predecessor. General Electric and Martin were still squabbling over who was responsible for the original explosion, and several botched static tests on the replacement rocket in early January did not augur well for the relaunch. Nonetheless, with Medaris and von Braun breathing down their necks, Hagen and the rest of the Vanguard bosses were determined not to lose their turn in Cape Canaveral's tight launch rotation schedule.

The weather, on Wednesday, January 22, boded equally ill, as the final, frenetic preparations got under way. "The night was miserably cold and wet," Stehling recalled. "With rain and hail alternating. Somehow, that night, the noise of the electric generators, the roaring of the gas compressors and the steady scurrying and shouting around the blockhouse, the squawking of the intercom boxes and the jangling of the telephones, the sizzling of the hamburgers in the Garbage [food dispensary] truck, the clicking of telemetry relays in the room, all seemed to be more discordant than usual, and we all had a premonition that the countdown would be unsuccessful."

Sure enough, with only four and a half minutes to launch on Thursday morning, a short circuit due to rain forced a postponement. The countdown clock was reset to 1:00 PM, and the rapidly evaporating liquid oxygen tanks were refilled for another try at 4:00 PM. More glitches pushed the liftoff time to 7:00 PM. Then with only nine minutes to go, a wall of ominous clouds rolled over the Cape. For an hour and a half, everyone waited for the front to pass. But it refused to budge. "Scrub," the safety officer finally ordered, to general groans and curses. They would have another go the next day. "By this time the field crew had the usual number of unshaven men with dark circles under their eyes, and that gastric acid bubble uprising," Stehling recalled. And there was the question of what to do with the fueled rocket. The liquid oxygen and especially the corrosive nitric acid in the second-stage tanks would wreak havoc on seals, valves, and plugs if left too long. Should the entire system be drained, which would require working through the night? No, said the contractors, the seals would hold another day. The red-eyed, nerve-racked navy engineers were dispatched to the Vanguard Motel in Cocoa Beach for a few hours of much-needed sleep.

At launchpad 26A, meanwhile, where dozens of army binoculars constantly trained on the competition, ABMA's anxious observers also took a welcome break from their nervous vigil. "Our people did not take kindly to the idea of sitting around twiddling their thumbs until Vanguard took off," Medaris later recalled. As everyone at the ABMA complex well knew, TV3BU had only another seventy-two-hour window within which to launch, and then it would be Explorer's turn. They were counting the hours.

The following day, with the sun beating down on the Florida coast and the all-clear signal given by the meteorologists, Vanguard got to within twenty-two hair-raising seconds of liftoff, when its umbilical cord stuck. It was supposed to release automatically just prior to flight, but now a technician in a cherry picker was sent out to disconnect the cord by hand. For the sixth time, the countdown clock was reset for another attempt three hours later. This time it got to T minus fourteen seconds, when a valve alarm sounded. The liquid oxygen had been in the tanks so long that it had frozen a valve in the open position. It would have to be drained, and another day would be lost, as the countdown was reset yet again, this time for 1:00 PM on Saturday. Now, Vanguard was truly running out of time. What's more, nitric acid had been sitting in the upper stage for several days now, and soon it would start eating away at the rocket's innards.

Saturday brought more delays and technical glitches, and at 11:00 PM the launch was postponed once again. The window had narrowed to less than twenty-four hours; Sunday, January 26, was TV3BU's last shot. At ABMA, engineers chain-smoked, gulped coffee by the gallon, and paced like expectant fathers. No one could concentrate on the Juno. The navy was getting closer and closer with every attempt. Eventually it would get its bird off the ground.

On Sunday, exhausted launch crews from both teams reassembled before noon. The three-hour countdown was slated to start at 1:00 PM, and Vanguard technicians in hard hats and gray coveralls were giving TV3BU a final once-over when a human shriek followed by an ear-piercing siren erupted from launchpad 18A. A worker was screaming in agony, holding his face. Brown fumes, the sign of an acid leak, were rising from the middle of the rocket. Firefighters were dispatched to douse the leak, while senior engineers ran to assess the damage. It was serious. Acid had burned its way into one of the motors. The entire second-stage engine would have to be replaced. A meeting of Vanguard's top personnel was hastily convened. They could scavenge a new motor from TV4, which was ready at the assembly hangar, but that would take time and would create new risks. There was really only one option, they realized with creeping dread: cancellation. "Above our meeting in the hangar hovered a ghostly consortium, von Braun and his ABMA group," a deflated Stehling recalled. "The Army rocket stood nearby, almost insolent. We had had it."

. . .

After years of rejection, months of upheaval, and four agonizing days, ABMA's fate was finally in its own hands. Now there were no more distractions, and Kurt Debus, von Braun's unflappable firing crew chief, could concentrate on getting Juno ready. Debus was in charge of all ABMA operations at the Cape. A Peenemünde alumnus and a veteran of over two hundred V-2 shoots, he was probably the most experienced launch master on the planet. Like Voskresenskiy with Korolev, he was one of the only people at ABMA who could override von Braun, and even Medaris deferred to his judgment once the countdown started. Quiet and unassuming, Debus spoke English with a heavy accent. He had an uncanny ability to parse the torrents of information that flooded the command center just prior to launch, and with his nerves of steel he had a seeming immunity to the adrenaline rush that sent everyone else's pulse racing so madly when rockets thundered to life. Debus did, however, have one failing, and it had almost prevented him from coming to the United States in 1945. Army investigators had classified him as "an ardent Nazi," who had "denounced his colleagues to the Gestapo." But such was Medaris's confidence in his firing chief, now a U.S. citizen, that von Braun would not even be present for the January 29 launch. Medaris wanted von Braun in Washington when Explorer went into orbit so that ABMA would be represented at the press conference the IGY committee was planning to hold the moment it got word that the mission was successful. Von Braun was unhappy with the arrangement, but Medaris insisted that he would be of far greater value in the capital, making sure the army got the credit it deserved. A great deal of future funding was riding on it.

With von Braun heading north to wage the public relations war in Washington, responsibility for Juno—though everyone at ABMA still called the rocket Jupiter C—rested entirely with Debus. Juno's first stage, the elongated Redstone, needed little prep work. All its components had been thoroughly tested in Huntsville. The carrier's upper sections, however, had to be carefully fitted together on site since they used solid propellant, a volatile mixture of polysulfide aluminum and ammonium perchlorate that was inherently unstable. Loading the eleven Sergeant rockets that powered the second stage was akin to handling live nitroglycerin charges, an operation best undertaken gingerly and not re-peated unnecessarily. A second, more complex phase of the assembly in-volved balancing the bundled rockets in the special spinning tub that was used to distribute thrust. All eleven motors had to push with the exact same strength at the exact same time for the second stage to work. The rotating platform, turning on its axis at 750 revolutions per minute, negated any irregularities in the individual rockets that might otherwise send the booster off course. But if it wasn't aligned, in perfect equilib-rium, it would vibrate and shake and tear the entire upper stage. Like a car mechanic balancing a wobbly wheel with tiny lead weights, Debus spent the better part of two days supervising minute calibrations on the spinning bucket.

Ernst Stuhlinger, meanwhile, tackled another critical task: a special timing device known as an apex predictor, which determined the precise moment when the second stage had to be fired to reach orbital velocity. Since there were no onboard computers in 1958 capable of quickly making such precise calculations, Stuhlinger would have to figure out the apex on the fly, using Doppler radar, telemetry readings, a slide rule, and some very fast calculations, and call the blockhouse to manually send a signal for the eleven Sergeants to simultaneously ignite. This was the trickiest part of the flight. A mathematical error, a downed phone line, or any other miscommunication could doom the entire mission. Korolev had gotten around the problem by having his giant core booster fire continuously, effectively one enormous stage. For Explorer, however, everything would come down to Stuhlinger, his slide rule, and his ability to speed-dial the command center.

Debus didn't like the arrangement. "Do you really want to rely on this alone?" he asked Stuhlinger, pointing to the intercom connection with the blockhouse. But Stuhlinger was ahead of him. He had set up his own ignition button as a backup in case his call couldn't get through. "I'll push it at the right moment," he promised. "Good," said a relieved Debus. "Good luck."

Luck, however, was not on ABMA's side, as January 29 rolled around and the jet stream howled in from the Atlantic with winds registering 175 knots at 45,000 feet, reaching 225 miles per hour in some pockets. Cape Canaveral's commander, General Donald Yates, had been Eisen-hower's meteorologist during the stormy Normandy invasion. He had brashly predicted before dawn on June 6, 1944, that the weather would clear, and Ike had gambled all on his being right. But now he shook his head with professional dismay. The jet stream would not shift, and Juno would not survive that kind of wind shear. With its elongated hull, retrofitted tanks, and added upper stages, it had been stretched to a perilously slender seventy feet, and the swirling crosswinds could twist it or snap it in half. For the sake of structural integrity, the launch would have to wait.

Now it was the army's turn to start sweating while the navy bided its time. The scrubbed TV3BU launch had been rescheduled for February 3, which meant that ABMA had to get its shot off by January 31 or lose its turn in the rotation. Vanguard still had priority over Explorer at Cape Canaveral, and since U.S. tracking stations could not juggle two satellites at once, a period of three days had to be left idle between rival attempts. So ABMA's window was now down to forty-eight hours.

The jet stream did not let up on Thursday, January 30. Despite the fact that at sea level only a gentle breeze ruffled the flags outside ABMA's assembly hangar, at 41,000 feet the winds raged at 205 miles per hour. High-altitude weather balloons were sent up every few hours to track the disruptive air currents, which showed some signs of subsiding by late afternoon. "What's happened? What are you going to do?" a helpless and clearly frustrated von Braun messaged frantically over the Pentagon's Teletype machine from Washington. Much like the Vanguard crew a few days earlier, Debus and Medaris now faced the dilemma of whether to fuel Juno. If the winds didn't die down, and the countdown was scrubbed, they would have to drain the rocket and replace all the seals rather than risk a repeat of TV3BU's corrosion problems. But if they were too cautious, they risked missing their opportunity, since the weather forecasts were growing increasingly optimistic. Debus decided to compromise: load the fuel but hold the liquid oxygen until the last moment. That would mean less work if they had to scrub.

A final set of weather balloons was released three hours before the scheduled 10:30 PM liftoff. As data floated back to receiving stations an hour later, the initial reports seemed promising. The liquid oxygen tankers were put on standby while Debus had the numbers sent to ABMA's Computation Lab in Huntsville for more detailed analyses.

"Highly marginal," the lab messaged at 9:20 PM. "We do not recommend that you try it."

Drain the rocket, Debus ordered, to collective groans. The engineers shook their heads in disbelief. To have come so far, to have battled back from the political brink so many times, only to bested by the wind. It was maddening. Unbelievable. The height of poetic injustice. And now they were down to their last shot, with Vanguard breathing down *their* neck.

Get some sleep, Medaris counseled his dejected crew. Tomorrow would be a long, hard day.

The first weather report on Friday, January 31, gave a little reason for hope. The high-altitude winds had tapered off slightly overnight but still gusted at 157 miles per hour. A Redstone would have no problem slamming through this turbulence, but the more fragile, overextended Juno could still sustain damage in such conditions. Medaris munched nervously on a ham and egg sandwich. "Everyone was going on sheer nerve," he recalled. "The men were tired. They had been working long and irregular hours, snatching sleep whenever they could."

Once again, liftoff was tentatively scheduled for 10:30 PM, and at 1:30 PM the countdown clock was set to T minus eight hours, leaving an hour leeway for unforeseeable delays. The wind was still not cooperating, and as he waited for weather updates Medaris chain-smoked and forced himself to catnap. By late afternoon decision time was approaching. Juno would have to start fueling soon. The highly noxious dimethylhydrazine von Braun had swapped for alcohol required special care, and technicians in hermetically sealed suits with integral breathing apparatuses needed extra time to load the toxic propellant. They would need to start the operation no later than 6:30 PM to be ready. It was do or die. For the umpteenth time, Medaris and Debus pored over the weather charts. Cape Canaveral's chief meteorologist, a twenty-four-year-old first lieutenant by the name of John Meisenheimer, predicted a shift in the jet stream by late evening, with winds declining to within acceptable norms. But not everyone agreed with the young lieutenant. If he was wrong, it could mean disaster and could set ABMA back cruelly. But if he was right and they didn't seize the opportunity, Vanguard would get another chance at making history. "Every man on the crew was conscious that the hopes of a Nation were riding with us," Medaris reflected. The hell with it. He would gamble the hopes of the nation, and the future of his five thousand employees, on the word of a twenty-four-year-old kid. Fuel the rocket, he ordered.

News that the launch was a go was quickly wired to Washington, where von Braun, Defense Secretary McElroy, Army Secretary Brucker, and the rest of the top brass descended on the Pentagon's main communications center to follow the final countdown on large-screen Teletype monitors with direct links to the Cape. President Eisenhower would not be present during the launch. He was at a golfing retreat in Augusta, Georgia, but Jim Hagerty, his press secretary, would keep him informed of the developments.

At 9:42 PM, a warning horn sounded on launchpad 26A, as a giant gantry crane was slowly pulled away and the gleaming white Juno was doused in the bluish embrace of powerful sodium searchlights. Vapors hissed and swirled from the missile and rose through cumulus clouds, which parted to reveal a bright waxing moon. A pebbled casing of ice encrusted Juno's midsection, reflecting the glare from a pair of red signal lights winking on the pad below.

"T minus fifty minutes," loudspeakers throughout Cape Canaveral blared, while at the Pentagon the VIPs read the Teletype. "The searchlights are going on and lighting up the vehicle," the Teletype relayed. "It's a beautiful sight."

"T minus fifteen," the countdown continued, and Medaris felt the bile rising in his stomach. "There is nothing that I have ever encountered to equal the feeling of suspended animation that comes during those last minutes," he later recalled. Soon, the automated firing sequence would commence, and there would be nothing to do but watch and wait and worry. "When the countdown reaches zero," Medaris teletyped Secretary Brucker, "the bird will not begin to rise immediately so don't be worried if we don't tell you it's on its way."

"T minus eight and counting. The blockhouse is buttoned up." The area around the launchpad was clear. Juno began powering up. Inside the rocket, motors whirred, valves opened and closed, and pressure started building up in the fuel pumps. The spinning bucket with the eleven second-stage Sergeants began to rotate, slowly at first, then faster, and faster still, until it was whizzing at 550 revolutions per minute and the entire missile hummed. T minus one hundred seconds.

Inside the sealed concrete blockhouse, the fifty-four systems engineers grew quiet, scanning their instruments for any signs of trouble as Debus ran through a final checklist. "T minus ten," he announced, his voice hoarse but calm. Just before 11:00 PM, the firing command was given, and the ignition switch was flipped. "Main stage!" For fourteen and three-quarters seconds, Juno remained on the pad, as flames tumbled beneath it, growing brighter and stronger, until the entire pad was shrouded in pink flaming dust. Then it moved. "It's lifting," the Pentagon's Teletype sang. "It's soaring beautifully."

In Washington and at the Cape, grown men danced and hugged and whooped like excited teenagers, shouting, "Go, baby! Go!" Overhead, the missile's red glow receded from view as it pierced the clouds, slashed through the edge of the jet stream, and rose toward the stratosphere. "It looks good. It looks good. Still going good," Medaris's information officer, Gordon Harris, dictated to the Teletype operator.

Five miles away from the blockhouse, in a small, equipment-laden cubicle at ABMA's noisy assembly hangar, Ernst Stuhlinger was also tracking the missile's progress, slide rule in hand. He'd practiced his calculations countless times, had the math down till it was almost second nature, but now that it mattered, there was no signal from him. More than six and a half minutes had elapsed since takeoff, and the Redstone main stage should have reached its apex by now. Where was the signal? T plus four hundred seconds. Juno was now 225 miles above the earth, in the nearly horizontal position needed to circumnavigate the globe. Still no signal. Something must be wrong. Get Stuhlinger on the line, Medaris frantically shouted, just as red panel lights flashed SECOND STAGE IGNITION. Stuhlinger had done it. But what if he had made a mistake? What if he was just a fraction of a second or degree off? Either way, they would know soon enough. The upper stages fired for only six and a half seconds each, in rapid automated succession. Thirteen agonizing seconds later, at six minutes and fifty-two seconds into the flight, relief swept the room. Another indicator light flashed. "It's in orbit," said a technician matter-of-factly. For a stunned instant, the blockhouse fell completely silent. Should they tell Washington? Harris asked, finally breaking the trance. "No," replied Medaris, with a smile. "Let 'em sweat a little."

But Medaris himself was not done sweating. Like Korolev nearly four months earlier, he too would now have a tense hour-and-a-half wait to see if Explorer had built up enough momentum to stay in orbit. Only when tracking stations on the West Coast picked up its signal after a complete revolution would he know for certain if the satellite was truly in orbit. "I'm out of coffee and running low on cigarettes," Army Secretary Brucker impatiently wired good-humoredly from Washington. "Send out for more and sweat it out with us," Medaris replied.

Von Braun, meanwhile, had taken out his own slide rule, calculating the estimated time Explorer would cross into signal range of Goldstone, the big tracking station in Earthquake Valley, California. It would take 106 minutes, he announced, at 12:41 AM.

At 12:40, William Pickering, the JPL chief responsible for Juno's upper stages, could no longer contain himself. "Do you hear her?" he asked the Moonwatch station in San Diego. "No, sir," came the reply.

"Do you hear her now?" he demanded two minutes later. Again, negative. "Why the hell don't you hear anything?" Pickering had lost his cool.

By now, everyone at the Pentagon and the Cape was becoming seriously concerned. Three, four, and then five minutes passed. Messages were sent to every station on the West Coast. Anything? Nothing. Explorer was now eight minutes overdue. Satellites simply weren't late. They were governed by immutable laws. Something must have gone wrong.

"Wernher," Secretary Brucker's tone turned suddenly icy, "what's happened?" Von Braun, for once, was at a loss for words. Just then, a message clattered off the Teletype. "They hear her, they hear her," a jubilant Pickering shouted. It was the Earthquake Valley station. "*Goldstone has the bird!*"

The United States of America had just entered the space age.

EPILOGUE

When told that Explorer was in orbit, Nikita Khrushchev reportedly shrugged. The race, he well knew, would no longer be so one-sided, now that a sleeping giant had been roused; and for the Soviet Union, it would be a contest of diminishing returns. But it did not matter.

Moscow had already scored its biggest gains by the time Juno soared into space, and those all-important early victories could never be pushed aside. In the eyes of the world, Sputnik made the Soviet Union a genuine superpower and America's equal, and this new status would persist regardless of whose future rockets flew farther, faster, or higher. The triumph was psychological and irreversible, and would endure until the Soviet Union itself disappeared into the dustbin of history one wintry day in 1991. Then, just as swiftly, Moscow's international image would revert to its pre-Sputnik reputation as a brutish and backward land.

Russia's dominance of the space race did not peak in 1958. Moscow was able to hold its lead for another three years, culminating with cosmonaut Yuri Gagarin's ride into orbit atop an R-7 on April 12, 1961. But by then the element of surprise was gone. America had learned not to underestimate its Communist adversary, and Washington had embarked on its own ambitious and long-term space program. The shock value was therefore not the same as with Sputnik, though the historical significance of Gagarin's flight was probably far greater and did indeed resonate throughout the rest of the world, particularly in developing countries.

The Soviet Union and Khrushchev, however, paid a steep military price for the early space triumphs. The R-7, for all its success as a heavy-lift vehicle and propaganda tool, was a failure as an ICBM. "It represented only a symbolic counter threat to the United States," Khrushchev later conceded, and was "reliable neither as a defensive nor offensive weapon." The very qualities that made it so adept at hurling large payloads into orbit rendered it almost useless as a fast-strike strategic weapon. It was just too big and unwieldy for war. It couldn't be hidden in silos, moved on mobile launchers, or adequately protected. It took too long to fuel, and the huge infrastructure it required made too inviting a target. Khrushchev exaggerated somewhat when he boasted that his factories would roll out R-7s like sausages. In the end, only seven were ever deployed, and only four launchpads were built capable of handling the mammoth missile, which meant that Moscow could realistically depend on getting off only four shots in a first-strike scenario. If the United States attacked first, only one or two of the big missiles might be fired in time. Or possibly none. Whichever the case, the R-7 would not keep America at bay. As a security shield, it was a failure.

Ultimately, the R-7 cost Russia its missile lead because Moscow had to go back to the drawing board to develop an entirely new ICBM. In that regard, Korolev's ploy to distract Khrushchev from the R-7's failings by launching satellites worked all too well. By the time the Soviet leader fully realized that he did not have a reliable intercontinental rocket, the United States was pumping billions of dollars into its neglected missile programs because of the Sputnik scare, rapidly making up the lost ground. Khrushchev's bluff ended up backfiring. When Dwight Eisenhower left office in January 1961, the United States had 160 operational Atlas ICBMs and nearly one hundred Thor and Jupiter IRBMs stationed in Europe to Moscow's meager reserve of four vulnerable R-7s. Ironically, the additional Jupiters that were produced to mollify Lyndon Johnson and the jittery American public after the Sputnik scare now haunted Washington. "It would have been better to dump them in the sea than dump them on our allies," Ike later commented. But the need to find a home for the superfluous missiles preoccupied Washington, exasperating superpower tensions. Great Britain had the Thors, which could hit only the Warsaw Pact countries and the westernmost parts of European Russia. But no frontline NATO allies wanted the Jupiters. In the end, Italy and Turkey reluctantly agreed to accept them in late 1959, and since they were geographically closer to Soviet borders, the Kremlin reacted furiously. "How would you like it if we had bases in Mexico and Canada?" fumed Khrushchev, angrily denouncing the deployment.

Tensions escalated further still, six months later, on May 1, 1960, when a U-2 was finally shot down over Soviet territory. Eisenhower, who was playing golf that day, as he had on the day of the very first U-2 mission four years earlier, would initially deny the incident, presuming that the pilot was dead and that the fragile aircraft had disintegrated, leaving little incriminating evidence. But as the Soviets kept pressing the issue, on May 5 the State Department would be forced to concede that a "civilian pilot of a weather-research plane" had indeed experienced problems with his oxygen supply over Turkey. "It is entirely possible that having a failure in the oxygen equipment, which would result in the pilot losing consciousness," the statement coyly reasoned, "the plane continued on automatic pilot for a considerable distance and accidentally violated Soviet airspace." A few days later, Washington would be further forced to eat its words when a beaming Khrushchev produced the CIA pilot Francis Gary Powers and the U-2's intact spy gear at an international press conference.

The incident caught Eisenhower in the devastatingly embarrassing lie that he had long predicted and feared, and spelled the end of manned reconnaissance flights into Russia. But just a few months later, on the same day that a Moscow court convicted Francis Gary Powers of spying, a new era of robotic, outer-space espionage began. On August 19, 1960, Richard Bissell's spy satellite successfully jettisoned its first batch of photographs of Soviet territory. Corona's film canister reentered the atmosphere off the shores of Hawaii, deployed its parachute, and was snagged in midair at 8,500 feet by grappling hooks attached to the front of a C-119 military plane.

And yet Bissell's triumph would be short-lived, as he was undone by the Bay of Pigs fiasco the following year. Begun under Eisenhower and executed under the new Kennedy administration, the botched attempt to topple the Cuban president Fidel Castro would prove even more embarrassing than the U-2 shoot-down. As the failed mission's architect and primary planner, Bissell—along with his patron and boss, Allen Dulles—would be forced out by the newly elected president, who would soon find himself baptized by rocket fire and international crisis.

"Those friggin missiles," as John F. Kennedy derisively referred to the Jupiters, finally caused Khrushchev to snap when they became operational in Turkey in late 1961. From their Turkish bases, they could hit military installations in the heart of the Soviet Union, effectively restoring the very same strategic imbalance that had prompted Moscow to build rockets in the first place. The net result was the Cuban missile crisis.

As it turned out, it would be a U-2, and not the top-secret Corona, that snapped the incriminating photographs of Soviet launchpad preparations on Castro's island that would spark the most dangerous showdown of the cold war. For Khrushchev, the attempt to station intermediate-range rockets on Cuban soil in the autumn of 1962 was a desperate gambit to redress the R-7's shortcomings. By placing smaller missiles within striking distance of America's shores, he sought to buy time for Yangel's R-16 to finish trials and go into mass production. The Soviet military, by then, had long switched its allegiance from Korolev's R-7 to rival missile designs, but the R-16 had suffered a series of developmental setbacks, including a catastrophic explosion of Glushko's acid propellant that claimed the lives of Marshal Nedelin and 112 other Soviet rocket scientists in 1960, when Nedelin disregarded Glushko's advice and ordered repairs performed on a fully fueled missile without draining it first. Only after the R-16 was fully ready, Khrushchev reasoned, would the balance of power be restored and security reestablished. To achieve that balance, he would risk confrontation. But in Cuba, Khrushchev made the wrong bet, and it would cost him the throne.

In the missile crisis it was Khrushchev who blinked first, promising to withdraw the IRBMs from Cuba. And even though Kennedy secretly agreed to remove the offending Jupiters from Turkey in exchange for the Soviet pullback, Khrushchev's days were numbered. In that sense, the R-7 was the vehicle through which Khrushchev's career soared and sank. Sputnik's glory cemented his grip on power. But when the R-7 proved a battlefield bust, and the missile foray into Cuba turned into a humiliating retreat, the resulting political wounds proved equally fatal. Rumblings of discontent started almost immediately in the Presidium. In 1963 they grew louder and bolder as Khrushchev's position further weakened due to his disastrous agricultural experiments. The ambitious farming reforms he had stubbornly rammed through the reluctant Presidium in 1957—the cultivation of millions of acres of "virgin lands" in Siberia and central Asia and the tripling of livestock quotas to overtake the United States in meat and dairy production—had completely collapsed. Not only were the thin-soiled Siberian fields ill suited for annual planting, but the quota system for increased meat and milk supplies served only to bankrupt many collective farms. To meet Khrushchev's unrealistic norms, wily farm bosses used funds allocated for machinery and buildings to buy cattle on the sly and then resold the animals to government agencies at a third of the price. The purchase and upkeep of tractors and combine harvesters were sacrificed for the paper gains, and the charade lasted just long enough to devastate the countryside. Far from overtaking the United States, as Khrushchev had boldly boasted in 1957, by early 1964 the gap had actually widened in America's favor. So poor were the harvests that the Soviet Union faced food shortages and, for the first time since the Second World War, rationing restrictions. Retail prices at official food stores rose 50 percent that year, many times more on the thriving black market, prompting protests from ordinary citizens and calls from indignant Central Committee factions for Khrushchev to answer for his "adventurism and irresponsibility."

In October 1964, while Khrushchev was vacationing at his dacha on the Black Sea, the decision was made to oust him. In the Central Committee, 197 of the 200 full members supported the secret vote of no confidence, selecting Leonid Brezhnev as first secretary in his place. Given a generous pension, a small staff, and the use of a large apartment and dacha, Khrushchev lived out his retirement peacefully. The virtually illiterate peasant who freed the Soviet Union from Stalin's terror and turned the USSR into an unlikely beacon of technological progress was the first leader in Russian history not to have died or been murdered in office. His son, Sergei, fulfilled his father's dream of earning a doctorate and became a rocket scientist. Eventually, he moved to America, where today he is a senior fellow at Brown University.

Under Brezhnev, cosmic conquests would lose priority and momentum in the Soviet Union. Moscow still aimed for the moon, the ultimate bragging ground, but the effort did not have the same intensity or urgency as the post-Sputnik rush to paint the heavens Red. Focus steadily shifted toward military missile expenditures, as funds dried up and the economic crisis that the CIA had long predicted worsened. In addition to its agricultural woes, Soviet industrial growth began to slow dramatically in the mid-1960s, actually contracting in many cases, and resources became increasingly scarce. With the milestones in the space race growing ever more ambitious and costly, the Kremlin's cautious new bosses preferred to spend on security rather than prestige.

In 1965, for the first time since its inception, OKB-1, Sergei Korolev's design bureau, began suffering budget shortfalls and cutbacks. The Chief Designer's star also began to fade. Khrushchev, his devoted patron, was gone, and Brezhnev did not have the same obsessive commitment to upstaging the Americans. Other missile makers were on the rise, landing big military orders, while Korolev's giant new Lunar rocket, the N-1, was mired in technical and financial problems. What's more, his quarreling with the imperious Valentin Glushko over the type of fuel to use on the 400-foot-tall N-1 had reached a point where the two were no longer on speaking terms, and not even Khrushchev could reconcile their differences. "I did everything I could to patch up their friendship," the Soviet leader recalled, "but my efforts were in vain." Worse, for the hypercompetitive Korolev, the United States was making very real strides with its proposed equivalent to N-1, the Saturn, and by the mid-1960s the Americans were poised to overtake him.

It was perhaps fortunate that Korolev died when he did, on January 11, 1966, before things began to unravel in earnest. Officially, the cause of death was complications during routine surgery to remove a tumor from his intestinal tract. But his colleagues said he worked himself to death. Korolev's heart gave out during the operation. It had always been weak and had grown weaker in the last few years of his life, forcing frequent hospital stays. In the end, the gulag and the relentless pressure that he imposed on himself finally took their toll. The hard-driving Chief Designer was fifty-nine years old.

Buried in the Kremlin wall near Lenin's tomb with all the pomp and ceremony of a national hero, Korolev was at last accorded the recognition he so richly deserved. He never realized his dream of putting a man on the moon, and as a weapon maker his most lasting contribution to the world's arsenals would not be the R-7, but the R-11, originally a small submarine-launched rocket whose land-based version would become more commonly known as the Scud. His enduring legacy, however, would be as a space pioneer, as the man who in total anonymity made America tremble. Had he lived another five years, perhaps history would have been rewritten; perhaps the hammer and sickle would have flown first on the moon, instead of the stars and stripes. But then again, the tide seemed to have turned by then, and it is not clear whether even the tenacious Chief Designer could have rescued the faltering Soviet space program. One will never know. What is certain is that Russia's moon dreams died with Sergei Pavlovich Korolev.

His influence on the United States, however, persists to this day. NASA, the institution created in the National Aeronautics and Space Act of 1958 as a direct response to Sputnik, landed a man on the moon, created the space shuttle, and is now probing farther and deeper into the cosmos. Millions of American students still benefit from the college loan programs started under the 1958 National Defense Education Act, which also revamped elementary and high school curricula with an emphasis on science and foreign languages to better compete with Soviet engineers. The Defense Reorganization Act of 1958 created the advanced military research agency that developed the Internet and countless other inventions that have transformed the daily lives of Americans.

Politically, Korolev and Khrushchev cast an equally long shadow across the United States. Without the "Sputnik Congress," and the Preparedness hearings that gave him such national exposure, Lyndon Johnson might not have won a spot on the 1960 Democratic ticket and ultimately may never have become president. NASA and the educational and military reforms of 1958 were all the creations of Johnson's hearings, and the perceptions of the "missile gap" that he first raised became a central issue in the 1960 election, not to mention a costly mainstay of defense expenditures for decades to come. Stuart Symington beat the missile gap drums almost as loudly and alarmingly as Johnson, but in the end this consummate cold warrior's presidential hopes withered because he refused to endorse escalating America's involvement in Vietnam. In the 1960s, he grew increasingly disillusioned with the CIA's covert activities in Indochina. Vietnam, he predicted, would prove an inescapable quagmire and ultimately a losing proposition. For his prescience, Symington was castigated as weak and soft on communism, and his career never recovered.

As the 1960 presidential poll approached, it looked as if Richard Nixon would at last have his just reward for all those painful and bitter dues he had paid at Ike's cold and distant shoulder. Personal relations between Eisenhower and his vice president never truly improved in the years after Sputnik, but Nixon, in preparation for the 1960 elections, was accorded a far greater role in Ike's second term. He traveled more frequently, especially after John Foster Dulles's death in 1959, and assumed many of the secretary of state's foreign policy duties. Some of the trips did not go smoothly. His infamous shouting match with Nikita Khrushchev at Moscow's international technology fair in July 1959 did little for the cause of improving superpower relations. Visiting a mock-up of an American kitchen displayed at the fair, the two leaders launched into an impromptu argument over missiles that ended with the red-faced first secretary jabbing a pudgy finger in the vice president's chest and growling: "You want to threaten? We will answer threats with threats."

Nixon's eventual opponent, John F. Kennedy, would make much of the unstatesmanlike buffoonery of the "kitchen debate," and he would owe a large debt to the continued fallout from Sputnik and the missile gap for his electoral victory the following year. But it would be Nixon who would preside over the White House when Kennedy's pledge to put a man on the moon was finally realized in 1969. Some would say this was fitting since he had advocated, as vice president, for Ike to shoot for the moon. But ultimately, would Kennedy have made the pledge, and would Neil Armstrong have taken his famously "small step" when he did, had Sergei Korolev not pitched Khrushchev the idea of a satellite? Perhaps not. The Chief Designer may be completely unknown to most Americans, yet his hidden hand has left indelible prints on the nation.

The *hidden hand* would become a term better associated with Eisenhower's detached style of leadership. Ike, after leaving office, did not enjoy rave reviews from historians, but by the mid-1980s scholarly esteem for Eisenhower had risen, a reevaluation that coincided, in part, with the declassification and release of many important documents, which revealed a vastly different man from the fuzzy, remote, and sometimes bumbling public persona. Behind closed doors, Eisenhower proved to be a far sharper figure, much more on top of issues than he publicly let on. It was the dichotomy that Richard Bissell had noticed, when he first assumed that John Foster Dulles ran the show, but after careful observation concluded that Ike was very much his own man. In hindsight, Ike's subdued response to Sputnik probably owed as much to his instinctive fear of the rise of the "military-industrial complex" as to his failing health and his longing for a peaceful retirement. His farewell address to the nation in January 1961 highlighted the danger of allowing the political and economic interests of military contractors and bureaucrats to hijack the national security agenda for their own gain. America did not heed his advice, however, and to this day trillions of tax dollars have been needlessly spent on unnecessary weapons systems that have not necessarily made the country safer.

Donald Quarles, Eisenhower's contentious point man on curbing runaway military spending, was to have succeeded Neil McElroy as secretary of defense in 1959. But he was felled by a massive heart attack that year, and his legacy as the man who oversaw America's earliest space and missile efforts would remain mixed at best.

Sputnik would taint Eisenhower's legacy as well. Contemporary revisionists are too charitable when they hail his passivity during Sputnik as exemplary. Leadership during times of crisis cannot be hidden or managed from a golf course. It must be assertive and overt. During the Sputnik crisis, Ike fell short on both counts, and populists like Lyndon Johnson stepped into the leadership vacuum.

Assertiveness was never a quality lacking in General Bruce Medaris, the closest thing America had to a Korolev. Though von Braun would get the credit for opening the heavens to the United States, it was really Medaris's iron will and stubborn refusal to yield to bureaucratic setbacks that lofted Juno into space. A lesser general might have accepted the Pentagon line and awaited orders, but had that happened ABMA would not have been prepared to respond as quickly as it did. He is the other unsung hero of this tale.

Since mavericks don't tend to last in large institutional settings, it came as no surprise that Medaris's military career ended shortly after

ABMA's 1958 triumph. Despite Explorer's political victory, the army lost the war for missile supremacy with the air force, and ABMA was gradually dismantled to make room for the new civilian space organization. Medaris vehemently opposed NASA's founding on the grounds that it would cannibalize his beloved agency, and his criticisms were so vocal that he had little choice but to resign his commission once von Braun's team was transferred to civilian control. In early 1960, Joseph P. Kennedy offered Medaris a job advising his son's presidential campaign on space issues, but he declined, wanting no part in politics. Instead, he accepted the presidency of the Lionel Corporation, a toy train maker with a defense contracting arm. A bout with cancer in the mid-1960s and a miraculous recovery left the devout former general convinced that he had been spared to fulfill a higher calling. He became a lay deacon in 1966, and four years later, at the age of sixty-eight, he was ordained an Episcopal priest. Father Bruce, as Medaris would be called in the final years of his life, passed away in 1990. He was buried with full honors at Arlington National Cemetery.

Wernher von Braun, of course, went on to become NASA's most famous founding scientist. He took America to the moon, became rich and respected, and fulfilled all his childhood dreams. His past, however, started catching up with him in the early 1970s. After *Paris Match*, the glossy French magazine, published a glowing article on the handsome space prophet, several readers wrote in to report that they recognized the man in the photographs. He was stouter and grayer than they remembered, but the burning eyes were unmistakably the same. The readers were survivors of Mittelwerk, former V-2 slave laborers, and the accounts they gave of von Braun differed starkly from the magazine's fawning profile. They claimed he had personally ordered prisoners executed for sabotage and was a war criminal who should face international tribunals.

Nothing came of the allegations, but von Braun spent the final years of his life defending his war record, and he died in 1977 under a growing cloud of suspicion. In 1984, another V-2 veteran, Arthur Rudolf, the designer of the Saturn V rocket that propelled the Apollo spacecraft to the moon, was quietly extradited to West Germany on identical charges. Von Braun unquestionably deserves a place in American history, but his true legacy remains murky.

As for the legacy of the first space race, the pioneering technology of the era is omnipresent in today's information age. Satellites govern virtually every aspect of modern life, from communications to credit card transactions to avoiding traffic jams using GPS receivers. The 2003 invasion of Iraq was the first military campaign in the history of warfare run almost entirely remotely, via satellite. Thanks to microchip implants, satellites monitor the whereabouts of wayward pets and track cargo and stolen vehicles. They transmit television signals to dish owners and cable operators and broadcast the rantings of radio personalities like Howard Stern.

Space is no longer the exclusive domain of superpowers, but increasingly it is a commercial battleground open to all who can afford it. In this new profit-driven arena, OKB-1, Korolev's old design bureau, is now called Russian Space Corporation Energya, and it supplies the boosters that orbit private U.S. satellites like DirectTV, beaming *The Sopranos* and National Football League packages to millions of American homes. Ironically, its partner in the rocket-for-hire venture is Boeing, the same company whose long-range bombers scared Nikita Khrushchev into founding OKB-1 to build the ICBM.

It's a fitting end to the Sputnik saga to see the former ideological rivals now working together in the common pursuit of market share. That, too, is a big part of the new wireless age that the launch of the world's first satellite made possible fifty years ago.

NOTES

Prologue

PAGE

1   Every second from now on meant 275 fewer pounds: See propellant use specifications at http://www.v2rocket.com/start/makeup/motor.html.
sixteen-ton Strabo crane: http://www.v2rocket.com/start/deployment/mobile operations/html.

2   the nearby horse-track oval outside suburban Wassenaar: http://www.v2rocket.com/start/deployment/denhaag.html.
trajectory traced by a billowy white vapor trail: Dieter K. Huzel, *Peenemünde to Canaveral* (Englewood Cliffs, N.J.: Prentice-Hall, 1962), p. 75.
producing a vein of jet exhaust gases at 4,802 degrees Fahrenheit: http://www.v2rocket.com/start/makeup/motor.html.
the second battery of the 485th Artillery Battalion: Frederick Ordway and Mitchell Sharpe, *The Rocket Team* (Burlington, Ontario: Apogee Books, 2003), p. 139.
body in a forty-five-degree inclination: Michael J. Neufeld, *The Rocket and the Reich: Peenemünde and the Coming of the Ballistic Missile Era* (Cambridge, Mass.: Harvard University Press, 1996), p. 97.

3   The gimbaled spinning wheels, rotating at 2,000 revolutions per minute: http://www.v2rocket.com/start/makeup/design.html.
Sixty-three seconds into its flight, the rocket ceased being a rocket: Huzel, *Peenemünde to Canaveral*, p. 75.
At an altitude of seventeen miles: Neufeld, *The Rocket and the Reich*, p. 98.
shell painted in a jagged camouflage scheme of signal white, earth gray, and olive green: http://www.v2rocket.com/start/makeup/markings.html.
moving at 3,500 miles per hour: Huzel, *Peenemünde to Canaveral*, p. 75.

4   Another ten seconds passed, and the rocket reached its apogee of fifty-two miles: http://www.v2rocket.com/start/deployment/timeline.html.

4   forward at nearly five times the speed of sound: http://www.centennialofflight
.gov/essay/EvolutionofTechnology/V-2/Tech26.htm.
The time was 6:41 PM: http://www.v2rocket.com/start/deployment/timeline
.html.
six-year-old John Clarke was freshening up for dinner: John Clarke interview
with BBC, September 7, 2004, at http://news.bbc.co.uk/1/hi/sci/tech/
3634212.stm.
the quarter-inch-thick sheet metal that encased it rose to 1,100 degrees:
http://www.v2rocket.com/start/makeup/design.html.

5   the V-2 slammed into Staveley Road at Mach 3: BBC, at http://news
.bbc.co.uk/1/hi/sci/tech/3634212.stm.
"The best way to describe it is television with the sound off ": Ibid.
"German science has once again demonstrated a malignant ingenuity":
William E. Burrows, *This New Ocean: The Story of the First Space Race* (New
York: Modern Library, 1999), p. 102.
Boris Chertok had no trouble finding the big brown brick building: To re-
create Chertok's experiences in Berlin, I have drawn on the English translation
of the first volume of his four-volume memoirs, *Rakety i Lyudi*: Boris Evsee-
vich Chertok, *Rockets and People* (Washington, D.C.: National Aeronautics and
Space Administration History Series, 2005).

8   "Oh, this German love for details and this exactness": Ibid., p. 221.
"We have every right to this": Ibid., p. 362.
"The thing that every laboratory needs the most": Ibid., p. 221.
"No, we no longer felt the hatred or thirst for revenge": Ibid.

9   "Occupation of German scientific and industrial establishments has revealed":
John Logsdon, ed., *Exploring the Unknown: SuDocNAS 1/1.21*, vol. 4 (Wash-
ington, D.C.: National Aeronautics and Space Administration, 1995), p. 33.
"The thinking of the scientific directors of this group is 25 years ahead": Ord-
way and Sharpe, *The Rocket Team*, p. 198.
He was a crack marksman, a recipient of the Knox artillery trophy and the
Distinguished Pistol Shot medal: See Toftoy's official biography at http://
www.redstone.army.mil/history/toftoy/memoir.html.

10   "It is no exaggeration to say that almost everything that [the class of] '26 has
done": Ibid.
"Hey Sarge, what do you think that odor could be?": Mary Nahas, *The Journey
of Private Galione: How America Became a Superpower* (Enumclaw, Wash.:
Pleasant Word Publishers, 2004), p. 276.

11   "They were gray in color, and they looked like skeletons": Ibid., p. 284.
"From where I was standing, I could see a hidden tunnel": Ibid.
The only unit that remotely fit that bill was the 144th Motor Vehicle Assem-
bly Company: http://www.v2rocket.com/start/chapters/mittel.html.
The Americans had hauled away one hundred intact rockets and had filled six-
teen Liberty Ships with 360 metric tons: Ibid.

12   "The problem is this": Quoted in Ordway and Sharpe, *The Rocket Team*,
p. 221.
nearly five hundred Russians, Poles, and Hungarian Jews: Nahas, *The Journey
of Private Galione*, p. 286.
"Most of their bodies have lost both trousers and shoes": Neufeld, *The Rocket
and the Reich*, p. 262.
5,789 V-2s produced at Mittelwerk: Ibid., p. 263.

13   "I know places where the SS hid the most secret V-2 equipment": Chertok, *Rockets and People*, p. 278.
      One of the biggest windfalls was dug out of a sand quarry in Lehesten: Ibid., p. 339.
      had been counterintuitively shortened and flared, creating a larger opening: http://www.centennialofflight.gov/essay/EvolutionofTechnology/V-2/Tech2 6.htm.

14   twenty thousand separate parts went into each V-2: Ernst Stuhlinger and Frederick I. Ordway, *Wernher von Braun: Crusader for Space* (Malabar, Fla.: Kreiger, 1994), p. 48.
      "down to the last screw": Chertok, *Rockets and People*, p. 282.
      "These documents were of inestimable value": Huzel, *Peenemünde to Canaveral*, p. 151.

15   "My sister goes to university wearing men's boots": Chertok, *Rockets and People*, p. 303.
      "We'd even hatched a plan to kidnap von Braun": Ibid.
      "One day, a group of men in American Army uniforms entered the schoolhouse in Witzenhausen": Ordway and Sharpe, *The Rocket Team*, p. 202.

16   "It would be an effective straitjacket for that noisy shopkeeper, Harry Truman": Michael Stoiko, *Soviet Rocketry* (New York: Holt, Rinehart and Winston, 1970), p. 73.

1: The Request

18   Red banners hailing the Twentieth Party Congress: To re-create the Presidium visit to NII-88, I relied on author interviews and e-mail exchanges with Sergei Khrushchev, as well as his memoir, *Nikita Khrushchev and the Creation of a Superpower* (University Park, Pa.: Penn State University Press, 2000).
      in the lane reserved exclusively for party high-ups, rode the three other Presidium members . . . Nikolai Bulganin, Lazar Kaganovich, and Vyacheslav Molotov: Author telephone interview with Sergei Khrushchev, November 22, 2005.
      pinned little notes with the Russian word *prick*: William Taubman, *Khrushchev: A Man and His Era* (New York: Norton, 2003), p. 232.

19   twenty-six-horsepower knockoffs of the 1938 Opel Kadett: http://www.autogallery.org.ru/m400.htm.
      row after row after row of mind-numbingly identical five-story apartment buildings: Nikita Khrushchev, *Khrushchev Remembers*, vol. 2, *The Last Testament*, edited by Strobe Talbott (Harmondsworth, Middlesex, England: Penguin, 1977), p. 141.

20   "little man with fat paws": Taubman, *Khrushchev*, p. 352.
      "After a year or two of school, I had learnt how to count to thirty": Ibid., p. 24.

21   "We weren't gentlemen . . . sense": Nikita Khrushchev, *Khrushchev Remembers*, edited by Strobe Talbott, 2nd edition (Boston: Little, Brown, 1974), p. 88.
      "He could barely hold a pencil in his calloused hand": Taubman, *Khrushchev*, p. 56.
      "My father felt this was the best, most honorable profession a man could have": Author telephone interview with Sergei Khrushchev, November 22, 2005.

21   "He wanted me to see the theories": Ibid.

22   "You see, I was studying to become a rocket scientist": Ibid.
      "Every villager dreamed of owning a pair of boots": Nikita Khrushchev, *Khrushchev Remembers*, edited by Strobe Talbott, 1st edition (Boston: Little, Brown, 1970), p. 266.

23   Khrushchev was unsettled by the rise to power of the Republican Party: Author telephone interview with Sergei Khrushchev, November 27, 2005.
      "the Soviets sought not a place in the sun": Peter Grose, *Gentleman Spy: The Life of Allen Dulles* (Amherst, Mass.: University of Massachusetts Press, 1996), p. 461.
      National Intelligence Estimate of September 15, 1954: Gerald K. Haines and Robert E. Leggett, eds., *CIA's Analysis of the Soviet Union, 1947–1990: A Documentary Collection* (Washington, D.C.: Center for the Study of Intelligence, Central Intelligence Agency, 2001), p. 49.
      "agonizing re-appraisal": Leonard Mosley, *Dulles: A Biography of Eleanor, Allen, and John Foster Dulles and Their Family Network* (New York: Dial Press, 1978), p. 307.

24   "liberate captive peoples" and "roll back": Ibid.
      prepare for "total war": Herman S. Wolk, "The New Look," *Air Force Magazine*, August 2003, http://www.afa.org/magazine/Aug2003/0803look.asp.
      "to create sufficient fear": Ibid.
      "We shall never be the aggressor": Ibid.
      2,280 atomic and thermonuclear bombs: David Alan Rosenberg, *Constraining Overkill: Contending Approaches to Nuclear Strategy, 1955–1965* (Washington, D.C.: Naval Historical Center, 2003), at http://www.history.navy.mil/colloquia/cch9b.html.
      the Strategic Air Command kept a third of its 1,200 B-47 long-range bombers: http://www.vectorsite.net/avb47_2.html.

25   Operation Power House: Ibid.
      Operation Home Run: James Bamford, *Body of Secrets: Anatomy of the Ultrasecret National Security Agency* (New York: Anchor Books, 2002), p. 36.
      "With a bit of luck, we could have started World War III": Thomas Coffey, *Iron Eagle: The Turbulent Life of General Curtis LeMay* (New York: Random House, 1986), p. 245.
      "Soviet leaders may have become convinced": Haines and Leggett, eds., *CIA's Analysis of the Soviet Union, 1947–1999*, p. 27.
      obliterating 118 of the 134 largest population and industrial centers: Rosenberg, *Constraining Overkill*, p. 8.
      The giant plane could carry 70,000 pounds of thermonuclear ordnance over a distance of 8,800 miles: http://www.af.mil/factsheets/factsheet.asp?fsID=83.
      an aging knockoff of the propeller-driven Boeing B-29 with a 2,900-mile range: http://www.globalsecurity.org/wmd/world/russia/tu-4.htm.

26   to get there visitors had to take a series of right turns: James Harford, *Korolev: How One Man Masterminded the Soviet Drive to Beat America to the Moon* (New York: John Wiley and Sons, 1997), p. 238.

27   only three people in the entire country would get one: Sergei Khrushchev, *Khrushchev on Khrushchev: An Inside Account of the Man and His Era* (New York: Little, Brown, 1990), p. 166.
      "it was always in a whisper": Author telephone interview with Sergei Khrushchev, November 22, 2005.

28   the president's advisers had spent much of that summit trying to figure out
      who was really running the show: Stephen Ambrose, *Eisenhower: Soldier and
      President* (New York: Simon and Schuster, 1990), p. 392.
      built in 1926 by the German firm of Rhein-Metall Borsig: Harford, *Korolev*,
      p. 78.
      "This is our past": Sergei Khrushchev, *Nikita Khrushchev*, p. 103.
29   "Father was no longer a novice when it came to missiles": Author telephone
      interview with Sergei Khrushchev, November 22, 2005.
      Beria, much like Hitler's secret police chief, Heinrich Himmler: Neufeld, *The
      Rocket and the Reich*, p. 214.
      "We gawked as if we were a bunch of sheep": Khrushchev, *Khrushchev Remem-
      bers*, 2nd edition, p. 46.
      except that it was nine feet longer, and of a slightly wider girth, which allowed
      it to carry extra fuel, doubling its range to nearly 400 miles: Asif A. Siddiqi,
      Sputnik and the Soviet Space Challenge (Gainesville, Fla.: University Press of
      Florida, 2000), pp. 62–63.
30   despite all the 15,000-ruble bonuses offered to captive German engineers:
      Chertok, *Rockets and People*, p. 366.
31   They had hovered over his deathbed like ghouls: Taubman, *Khrushchev*, p. 238.
      "If now, at the fountain of communist wisdom": Haines and Leggett, eds.,
      CIA's Analysis of the Soviet Union, 1947–1990, p. 50.
      7 million Soviet citizens: Roy A. Medvedev and Zhores A. Medvedev,
      Khrushchev: The Years in Power (New York: Norton, 1978), p. 15.
      "Don't you see what will happen?": Sergei Khrushchev, *Nikita Khrushchev*, p. 95.
      and engulfed 18 million lives: Anne Applebaum, *Gulag: A History* (New York:
      Anchor Books, 2003), p. 580.
32   "They had to be isolated": Taubman, *Khrushchev*, p. 216.
      "All it took was an instant": Ibid., p. 202.
33   "A change from violence to diplomacy": Haines and Leggett, eds., *CIA's
      Analysis of the Soviet Union, 1947–1990*, p. 52.
      1,548,366 arrested: Sergei Khrushchev, *Nikita Khrushchev*, p. 98.
      nearly always ended in death: Medvedev and Medvedev, *Khrushchev*, p. 20.
      "He was taken to Lefortovo prison, interrogated, beaten": Harford, *Korolev*,
      p. 52.
34   This time it flew 390 miles: Siddiqi, *Sputnik and the Soviet Space Challenge*, pp.
      98–99.
      "The construction looked utterly incapable of flight": Sergei Khrushchev,
      Nikita Khrushchev, p. 103.
      The engine, an RD-103 designed by Glushko: Ibid., pp. 100–101.
35   "Korolev walked over to a map of Europe": Ibid., pp. 103–4.
37   Khrushchev had approved the cinematic thaw: Pavel Loungine, director, *The
      Moscow Skyscraper*, British-French documentary film (Paris: Roche Produc-
      tions, 2004).
      Yields were so low that in 1953 per capita grain production: Medvedev and
      Medvedev, *Khrushchev*, p. 58.
38   which required relocating three hundred thousand farmworkers: Ibid.
      14 and 20 percent of the Soviet economy, compared to 9 percent for the
      United States: Haines and Leggett, eds., *CIA's Analysis of the Soviet Union,
      1947–1990*, p. 175.
39   "the striking re-allocation of expenditures": Ibid., p. 187.

39   "I was amazed": Author telephone interview with Sergei Khrushchev, November 27, 2005.

40   where the thermonuclear warhead would sit: Siddiqi, *Sputnik and the Soviet Space Challenge*, pp. 128–29.

41   "small-time cattle dealer": Taubman, *Khrushchev*, p. 267.
      "Their relations had become tense": Author telephone interview with Sergei Khrushchev, November 27, 2005.

42   "Comrade Khrushchev carries out his work . . . intensively, steadfastly, actively and enterprisingly": Ibid., p. 269.
      "led us to a stand occupying a modest place in the corner": Sergei Khrushchev, *Nikita Khrushchev*, p. 110.

43   Decrees had been signed advocating the "artificial moon": Siddiqi, *Sputnik and the Soviet Space Challenge*, pp. 145, 149.
      "You needed the constant support of power": Author telephone interview with Sergei Khrushchev, November 27, 2005.

44   "The Americans have taken a wrong turn": Sergei Khrushchev, *Nikita Khrushchev*, p. 111.
      "It seemed as if he was still debating the matter": Ibid.
      "If the main task doesn't suffer, do it": Matt Bille and Erika Lishock, *The First Space Race: Launching the World's First Satellites* (College Station, Tex.: Texas A&M University Press, 2004), p. 63.

2: Jet Power

46   "politeness is nice": Gordon Harris, *A New Command: The Story of a General Who Became a Priest* (Plainfield, N.J.: Logos International, 1976), p. 116.
      "Didn't you see the speed limit sign back there?": Ibid., p. 146.

47   so strongly favored the young air force that it now swallowed forty-six cents: Colonel Mike Worden, *Rise of the Fighter Generals* (Maxwell Air Force Base, Ala.: Air University Press, 1998), p. 89. Also at http://aupress.au.af.mil/Books/Worden/Worden.pdf.
      "You are aggressive. Some would say to a fault": Harris, *A New Command*, p. 127.
      did not enjoy "a great reputation": John B. Medaris, *Countdown for Decision* (New York: G. P. Putnam's Sons, 1960), p. 104.

49   Of the 7,920,000 automobiles sold by Detroit in 1955: Ambrose, *Eisenhower*, p. 386.
      his salary was diminishing from $566,200 to $22,500: *Time*, December 1, 1952, at http://www.time.com/time/magazine/printout/0,8816,817434,00.html.
      "what was good for the country was good for General Motors": http://www.defenselink.mil/specials/secdef_histories/bios/wilson.htm.

50   "kennel dogs" and "worry about what makes the grass green": *Time*, October 6, 1961, at http://www.time.com/time/magazine/printout/0,8816,827790,00.html.
      "Damn it, how in the hell did a man as shallow": William Bragg Ewald Jr., *Eisenhower: The President* (Englewood, N.J.: Prentice Hall, 1981), p. 192.
      "In his field, he is a competent man": Robert H. Ferrel, ed., *The Eisenhower Diaries* (New York: Norton, 1981), p. 237.
      military spending still ate up more than half the federal budget: http://www.army.mil/cmh-pg/books/amh/AMH-26.htm.

50   the New Look Defense Policy: Herman S. Wolk, "The New Look," *Air Force Magazine*, August 2003, http://www.afa.org/magazine/aug2002/0803look.asp.

51   "ambassadors to unfriendly nations": Medaris, *Countdown for Decision*, p. 104. The Redstone was a heavy-lift tactical missile capable of flinging a 3,500-pound nuclear warhead 200 miles: http://www.redstone.army.mil/cron2a.html; also http://www.boeing.com/history/bna/redstone.htm.

52   "to inflict very great, even decisive, damage": Ferrel, ed., *Eisenhower Diaries*, p. 324.
      "The world in arms is not spending money alone": Walter A. McDougall, *The Heavens and the Earth: A Political History of the Space Age* (New York: Basic Books, 1985), p. 114.
      the air force had spent a mere $14 million developing its ICBM by 1954: Ibid., p. 104.

53   increased missile spending to $550 million in 1955: Ambrose, *Eisenhower*, p. 41.
      far below the $7.5 billion earmarked for beefing up the bomber fleet: Worden, *Rise of the Fighter Generals*, p. 187.
      Did the accelerated spending "go far enough?": "Discussion of the 258th Meeting of the National Security Council, Thursday, September 8, 1955," 15 September 1955, NSC series, box 7, Eisenhower Papers, 1953–1961 (Ann Whitman file), Dwight D. Eisenhower Library, Abilene, Kansas.
      "I was always convinced that you would move ahead to the top": Christopher Matthews, *Kennedy and Nixon: The Rivalry That Shaped Postwar America* (New York: Simon and Schuster, 1996), p. 79.

54   "his lack of maturity": Ibid., p. 105.
      "You're my boy": Andrew J. Dunar, *America in the Fifties* (Syracuse, N.Y.: Syracuse University Press, 2006), p. 97.
      "Mr. Wilson started to ask some odd questions": Medaris, *Countdown for Decision*, p. 107.
      "it was the first of many shocks to come": Ibid.

55   John Foster Dulles, it was decided, would speak for the administration: Ambrose, *Eisenhower*, pp. 396–97.

56   "I will never answer another question on this subject": Matthews, *Kennedy and Nixon*, p. 104.
      "Every piece of scientific evidence that we have indicates": Ibid., p. 113.

57   "You said something about 'being afraid' ": Walter J. Boyne, "Stuart Symington," *Air Force Magazine*, February 1999, http://www.afa.org/magazine/feb1999/0299symington.asp.
      "He is a formidable-looking figure": *Time*, January 19, 1948, at http://www.time.com/time/magazine/printout/0,8816,779517,00.html.
      "We feel, with deep conviction": Worden, *Rise of the Fighter Generals*, p. 42.
      The USSR would have four hundred Bisons and three hundred Bears: John Prados, *The Soviet Estimate: US Intelligence Analysis and Russian Military Strength* (New York: Dial Press, 1982), p. 45.
      "We believe that in the future": Worden, *Rise of the Fighter Generals*, p. 87.

58   It revealed a high numeric series, which implied a vast production line: Grose, *Gentleman Spy*, p. 402.
      "unconstitutionally contradicting patriots": Prados, *The Soviet Estimate*, p. 44.
      only 85 of the 700 new bombers projected by air force intelligence: http://www.thebulletin.org/articles.php?art_ofn=ja01staff.
      "You'll never get court-martialed": Christopher Simpson. *Blowback: America's*

Recruitment of Nazis and Its Effects on the Cold War (New York: Weidenfeld and Nicolson, 1988), p. 64.

58   many of the air force officers who provided the testimony and information for the hearings were promoted: Prados, *The Soviet Estimate*, p. 50.

59   "frequent changes of scene and recreation": Fred I. Greenstein, *The Hidden-Hand Presidency: Eisenhower as Leader* (New York: Basic Books, 1982), p. 8.
      "the much publicized golfing trips, the working vacations, and even the Wild West stories": Ibid., p. 39.

3: Trials and Errors

60   the continuous blaring of car horns: To re-create the Georgian uprising, I relied on Sergei Stanikov's eyewitness account, which was published in the Russian journal *Istochnik*, no. 6, 1995. An English translation by Tahir Asghar is available at http://www.revolutionarydemocracy.org/rdv5n2/Georgia.htm.
      one of the first nations on earth to have adopted Christianity in A.D. 337: http://www.parliament.ge/pages/archive_en/history/his2.html.

61   "Great Son of the Georgian People" had been denigrated: Medvedev and Medvedev, *Khrushchev*, p. 70.
      "A meeting was held at 4 o'clock in which I was present": http://www.revolutionarydemocracy.org/rdv5n2/Georgia.htm.

62   they were currently reading the manuscript of a young writer named Boris Pasternak: Taubman, *Khrushchev*, p. 385.

64   nine protesters were officially pronounced dead: Sergei Khrushchev, *Nikita Khrushchev*, p. 164.

65   Khrushchev . . . could easily "be beguiled": Taubman, *Khrushchev*, p. 131.
      Glushko was elegant and regal: V. F. Rakhmanin, ed., *Odnazhy I Navsegda: Dokumenty I Lyudi o sozdatelye raketnykh dvigateley Valentnye Petrovichye Glushko* (Moscow: Mashinostroyenye, 1998), p. 341.
      Korolev, on the other hand, never wore a tie unless he had to: A. V. Ishlinskiy, ed., *Akademik S. P. Korolev: Uchonie, Inzhenier, Chelovek* (Moscow: Nauka, 1986), p. 107.

66   "He ate very quickly": Arkady Ostashov, Yuri Mozhorin, ed., *Nachalo Kosmichiskoy Eri: Vospominaniya Veteranov Raketno-Kosmicheskoy Tekniki*, vol. 2 (Moscow: RNITSKD, 1994), p. 44.
      the rasp of the needle on the gramophone: Deborah Cadbury, *Space Race: The Epic Battle Between America and the Soviet Union for the Dominion of Space* (New York: HarperCollins, 2006), p. 78.
      "Glushko gave testimonies about my alleged membership of anti-Soviet organizations": Ibid., p. 85.

67   "For the sake of my sole son": Harford, *Korolev*, p. 50.
      Glushko had also been in the camps: Rakhmanin, ed., *Odnazhy I Navsegda*, pp. 424–33.
      Korolev had engaged in a long-running affair with Glushko's sister-in-law: Cadbury, *Space Race*, p. 89.

68   "Sergei Pavlovich, your Horche is beautiful, but it's not a fighter plane": Chertok, *Rockets and People*, pp. 350–51.
      "I'm not afraid of anyone in the whole wide world": Ibid.

69   The solution was the RD-107: Valentin Glushko, *Raketnie Dvigateli GDL-OKB* (Moscow: Izdatelstvo Agentsva Pechatiy Novosti, 1975), pp. 328–29.

69   But if the central R-7 engine block was designed to operate longer, the four peripheral boosters could be jettisoned: *Novosti Kosmonavtiki*, vol. 15, no. 7 (August 2005), pp. 67–69.

70   Korolev favored using small gimbaled thrusters: Ibid., vol. 15, no. 8 (July 2005), pp. 56–59.

But Glushko was violently opposed to the idea: Yuri Semenov, ed., *Raketno Kosmicheskaya Korporatsiya Energiya Imeni S. P. Koroleva* (Korolev: RKK Energiya, 1996), p. 75.

could withstand the heat that would be generated during the 24,000-feet-per-second atmospheric reentry: Ivan Prudnikov, *Aviatsiya I Kosmonavtika*, vol. 1, no. 2 (1994), p. 39.

postimpulse boost: Timofei Varfolomeyev, *Space Flight Magazine* (UK), August 1995, p. 262.

71   "I've been sent the protocol of the latest tests": Cadbury, *Space Race*, p. 124.

how many pounds of thrust were produced per each pound of propellant consumed per second: Wernher von Braun et al., *Space Travel: A History* (New York: Harper and Row, 1975), p. 136.

But Glushko's engines had come up short, at 239 and 303.1 respectively: Georgiy Vetrov, ed., *S. P. Korolev I Evo Dela: Svet I Teni v Istorii Kosmonavtiki* (Moscow: Nauka, 1998), pp. 220–21.

72   "At present time, we are completing static testing of the rocket": Ibid., p. 369.

WE WANT BREAD, FREEDOM, AND TRUTH: Filip Lesniak, *Biuletyn Instituty Pamiecy Narodowek*, at http://www.ipn.gov.pl/biuletyn5_12.htm.

73   thirteen-year-old Romek Strzalkowski fell dead: Polish Academic Information Center, University of Buffalo, at http://www.info-poland.buffalo.edu/exhib/Poznan/june1956.html.

"The Poles were vilifying the Soviet Union": Nikita Khrushchev, *Khrushchev Remembers*, 1st edition, p. 198.

74   "From the airport we went to": Ibid., p. 200.

"To all those suffering under communist slavery": http://www.historylearn ingsite.co.uk/hungary_1956.htm.

eighty were killed: http://www.info-poland.buffalo.edu/exhib/Poznan/june 1956.html.

75   "The Soviet government is prepared to enter into the appropriate negotiations": http://www.gwu.edu/~nsarchiv/NSAEBB/NSAEBB76/doc6.pdf.

"This utterance is one of the most significant to come out": Grose, *Gentleman Spy*, p. 348.

defiantly summoned the Soviet ambassador, Yuri Andropov: http://www.gwu .edu/~nsarchiv/NSAEBB/NSAEBB76/doc7.pdf.

"We have no choice": Minutes of October 31, 1956, Presidium meeting, at http://www.gwu.edu/~nsarchiv/NSAEBB/NSAEBB76/doc6.pdf.

76   "Bombs, by God!": Grose, *Gentleman Spy*, p. 349.

"The Soviet Air Force has bombed the Hungarian capital": See online transcript at http://news.bbc.co.uk/onthisday/hi/dates/stories/november/4/newsid_2739000/2739039.stm.

77   "Khrushchev's days are numbered": Grose, *Gentleman Spy*, p. 337.

4: Tomorrowland

78   As the incumbent, Ike was able to rise above the political fray: Dunar, *America in the Fifties*, p. 123.

79   "In regard to the Intermediate Range Ballistic Missiles": Erik Bergaust, *Wernher von Braun* (Harrisburg, Pa.: Stackpole Books, 1976), p. 245.
       "it will be better for the country": Ibid.

80   "If we let down our standards to speed production": Harris, *A New Command*, p. 134.
       "The lack of a sound, experienced, military-technical organization": Medaris, *Countdown for Decision*, p. 57.

81   "In all honesty, I do not think": Ibid., p. 60.
       "Can you picture a war": Ambrose, *Eisenhower*, p. 410.
       "somewhat distorted and exaggerated picture": Worden, *Rise of the Fighter Generals*, p. 82.
       "satisfactory state of reliability": Ibid.
       "I don't know how to show . . . teeth with a missile": Ibid., p. 84.

82   "We see too few examples of really creative": Ibid., p. 80.
       "The aircraft industry, and particularly the Douglas Aircraft Co.": Bergaust, *Wernher von Braun*, pp. 252–53.

83   "It was not a big decision": Ernst Stuhlinger, December 8, 1997, interview with Michelle Kelly for the Johnson Space Center Oral History Project, NASA Oral History Transcript at http://www.jsc.nasa.gov/history/oral_histories/participants.htm.
       "Screen them for being Nazis?": Dennis Piszkiewicz, *Wernher von Braun: The Man Who Sold the Moon* (Westport, Conn.: Praeger, 1998), p. 41.

84   impregnable blackness: Stuhlinger and Ordway, *Wernher von Braun*, p. 70.
       "To my continental eyes": Piszkiewicz, *Wernher von Braun*, p. 8.
       Fort Bliss, an old cavalry outpost built around the rough adobe walls: http//www.bliss.army.mil/museum/fortblisstexas.htm.

85   "A line of waiters in black suits, white shirts, and bow-ties: Chertok, *Rockets and People*, p. 241.
       "German Scientist Says American Cooking Tasteless": *El Paso Times*, December 4, 1946.
       "We hold these individuals to be potentially dangerous": James McGovern, *Crossbow and Overcast* (New York: William Morrow, 1964), p. 247.
       "I never thought we were so poor mentally": Paul Dickson, *Sputnik: The Shock of the Century* (New York: Berkley, 2001), p. 61.

86   As wards of the army: Huzel, *Peenemünde to Canaveral*, p. 215.
       "Daily life was quite regulated": Stuhlinger and Ordway, *Wernher von Braun*, p. 77.
       "SPECIAL WAR DEPARTMENT EMPLOYEE. In the event that this card is presented off a military reservation": Huzel, *Peenemünde to Canaveral*, p. 223.
       they catch a screening of *Zorro* at the Palace Theatre, go shopping at the Popular Dry Goods Company: http://www.elpasotexas.gov/walkingtours.

87   playing cello in a string quartet of rocket scientists: Stuhlinger and Ordway, *Wernher von Braun*, p. 11.
       "Prisoners of peace": Ordway and Sharpe, *The Rocket Team*, p. 237.

87   "Frankly we were disappointed": Daniel Lang, "A Romantic Urge," *New Yorker*, November 7, 1951, p. 89.
88   "We'll put you on ice": Stuhlinger, NASA 1997 Oral History.
       "control of German individuals who might contribute": http://www.milnet .com/cia/nazi-gold/art04.html.
       "threat to world security": Lieutenant David Akens, *Army Ballistic Missile Agency Historical Monograph* (Huntsville, Ala.: Redstone Arsenal, 1958), p. 25, at http://www.redstone.army.mil/history.
       "We were distrusted aliens": Bob Ward, *Dr. Space: The Life of Wernher von Braun* (Annapolis, Md.: Naval Institute Press, 2005), p. 67.
89   when three hundred thousand sorties were flown: http://www.usafe.af.mil/ berlin/quickfax.htm.
       "when we might have completely destroyed Russia and not even skinned our elbows": David Halberstam, *The Fifties* (New York: Fawcett Columbine, 1993), p. 25.
       "greatest act of stupidity of the McCarthyist period": Dickson, *Sputnik*, p. 138.
       The historic hamlet was home to fifteen thousand genteel southerners: http://www.redstone.army.mil/history/cron2a.html.
90   "We had some concerns here": Ward, *Dr. Space*, p. 77.
       The fledgling ABC network was backing the venture with $4.5 million in loan guarantees: Piszkiewicz, *Wernher von Braun*, p. 84.
91   whose hourly pay in 1954 had just been increased from seventy cents to a dollar: Ambrose, *Eisenhower*, p. 386.
       "In our modern world": March 9, 1955, "Tomorrowland" telecast, *Walt Disney Treasures*, Buena Vista Home Entertainment, Burbank, Calif., 2004, stock no. 31749.
       The show attracted 42 million viewers: J. P. Telotte, "Disney in Science Fiction Land," *Journal of Popular Film and Television*, Spring 2005, http://www .finarticles.com/p/articles/mi_m0412/is_1_33/ai_n13717415.
92   to launch a satellite using a modified Redstone missile for less than $100,000: McDougall, *The Heavens and the Earth*, p. 119.
       "The atmosphere of the earth acts as a huge shield": http://www.history.nasa .gov/sputnik/chapter2.html.
93   "I am impressed by the costly consequences": Ibid.
94   "It must be restated": Haines and Leggett, eds., *CIA's Analysis of the Soviet Union, 1947–1990*, p. 59.
       "I wouldn't care if they did": Bille and Lishock, *The First Space Race*, p. 51.

5: Desert Fires

95   wads of rubles that Korolev kept in an office safe: Harford, *Korolev*, p. 4.
96   building nine tracking stations deep in the Kazakh desert over the first 500 miles: Siddiqi, *Sputnik and the Soviet Space Challenge*, pp. 156–57.
       A gigantic vise with collapsible jaws, pivots, and counterweights: Semenov, ed., *Raketno Kosmicheskaya Korporatsiya Energiya*, pp. 76–77.
       a combination of silica and asbestum with textalyte: Prudniko, *Aviatsiya I Kosmonavtika*, p. 39.

97   The pace of construction at Tyura-Tam had been so frenetic: Baikonur
      Cosmodrome   Foundation: http://www.russianspaceweb.com/baikonur_
      foundation.html.
      a fire alarm was inadvertently triggered, setting off sprinklers: Siddiqi, *Sputnik
      and the Soviet Space Challenge*, p. 157.
98   "What the hell are you doing?": Boris Chertok, *Rakety I Lyudi*, vol. 2
      (Moscow: Mashinostroyeniye, 1996), pp. 185–86.
      "Get him out of here": Yuri Mozzhorin, ed., *Dorogi v Kosmos*, vol. 1 (Moscow:
      MAI, 1992), p. 114.
      "That was Korolev": Sergei Khrushchev, *Nikita Khrushchev*, p. 221.
99   "Give me a crane, some cash": Mozzhorin, ed., *Nachalo Kosmichiskoy Eri*, p. 67.
      "Take it away": Chertok, *Rakety I Lyudi*, vol. 2, p. 177.
      Korolev developed strep throat and had to take frequent penicillin shots: Sid-
      diqi, *Sputnik and the Soviet Space Challenge*, p. 159.
      "We are working under a great strain": Cadbury, *Space Race*, p. 159.
100   "We are criminals": Vladimir Parashkov and Konstantin Gerchik, eds., *Niez-
      abivayemi Bajkanur* (Moscow: Rosijskoye Kosmichiskoye Agenstvo, 1998),
      p. 107.
      "What can they do to us?": Chertok, *Rakety I Lyudi*, vol. 2, p. 183.
101   "He was a brilliant scientist": Author telephone interview with Sergei
      Khrushchev, February 15, 2006.
      They were "worthless": Cadbury, *Space Race*, p. 159.
      One general had famously groused: Author interview with Peter Gorin,
      Williamsburg, Virginia, March 6, 2006.
      "You can't count on Malinovsky": Chertok, *Rakety I Lyudi*, vol. 2, p. 184.
102   Korolev was fixated on the notion: Harford, *Korolev*, p. 3.
      "And you and your rocket": The following exchange is quoted in Yaroslav
      Galovanov, *Korolev: Fakti I Mythi* (Moscow: Nauka, 1994), pp. 502–8.
103   "When things are going badly": Siddiqi, *Sputnik and the Soviet Space Challenge*,
      p. 159.
      "Sergei was about three": Ishlinskiy, ed., *Akademik S. P. Korolev*, p. 29.
104   As a side business, the family had a small but highly successful brine opera-
      tion: Natalia Koroleva, *Otets*, vol. 1 (Moscow: Nauka, 2001), p. 27.
      "He didn't have any friends of his own age": Harford, *Korolev*, p. 19.
      that his estranged father, whom he was not permitted to see, would try to kid-
      nap him: Koroleva, *Otets*, vol. 1, p. 65.
      "I felt I needed to keep him at home": Ishlinskiy, ed., *Akademik S. P. Korolev*,
      p. 30.
      Korolev built giant dollhouses and cried frequently: Koroleva, *Otets*, vol. 1,
      p. 62.
      "A poster appeared": Ishlinskiy, ed., *Akademik S. P. Korolev*, p. 29.
      No one in Nezhin had ever seen an airplane before: Koroleva, *Otets*, vol. 1,
      p. 68.
105   "Mother, can you give me two new bed-sheets": Ishlinskiy, ed., *Akademik S. P.
      Korolev*, p. 30.
      "Hunger, chaos": Ibid.
106   "hang onto the barbed wire": Ibid.
      "He was not interested in small talk": Koroleva, *Otets*, vol. 1, p. 116.
      1922 daily planner: Ibid., p. 134.
      "Oh Mother, if you could only see": Ibid., p. 121.

107   "That was the definitive moment": Ishlinskiy, ed., *Akademik S. P. Korolev*, p. 32.
Polytechnical Institute of Kiev, which produced such graduates as Igor Sikorsky: Harford, *Korolev*, p. 26.
"To my dear friend Piotr Frolov": Koroleva, *Otets*, vol. 1, p. 325.
when he met two rocket enthusiasts, Friedrich Tsander and Mikhail Tikhonravov: Georgy Vetrov, *Korolev I Kosmonavtika: Pervye Shagi* (Moscow: Nauka,
1994), p. 33.

108   "the air of a man who had already sampled the mysteries": Cadbury, *Space
Race*, p. 124.
Tikhonravov would coin the term *cosmonaut*: V. Davydova, "100-letie So Dnya
Rozhdeniya M. K. Tikhonravovna," *Novosti Kosmonavtiki*, October 2000, p. 61.
Korolev hit upon the idea of grafting it to a tailless, trapezoidal glider: Vetrov,
ed., *S. P. Korolev I Evo Delo*, p. 33.

109   "But Father, how could your plane land": Natalia Koroleva televised interview
for the documentary film *The Secret Designer*, Ryan Productions., Toronto,
1994.
"Nikita, come to the Kremlin": Sergei Khrushchev, *Nikita Khrushchev*, pp.
233–34.
"We will bury you": Medvedev and Medvedev, *Khrushchev*, p. 75.

110   The attack on Nikita Khrushchev began: The account of the attempted coup
against Khrushchev is drawn from several interviews with Sergei Khrushchev
in early 2006, his memoir of his father, Nikita Khrushchev's own multivolume
memoirs, and William Taubman's *Khrushchev*.

111   a cow "knocking about the whole country": Taubman, *Khrushchev*, p. 318.
"Leonid Ilyich barely had time to utter the first words": Sergei Khrushchev,
Khrushchev, p. 237.
"You've become the expert on everything": Author telephone interview with
Sergei Khrushchev, February 10, 2006.
"They couldn't, as long as Father retained the loyalty of two key people": Ibid.

112   signed 38,679 execution orders: Taubman, *Khrushchev*, p. 320.
"roared like an African lion": Ibid., p. 323.

113   "Okay, Boris, you continue playing sick": Chertok, *Rakety I Lyudi*, vol. 2, p. 190.

6: Pictures in Black and White

115   On the morning of August 28, 1957: A list of all U-2 Soviet overflights can be
found at http://www.spyflight.co.uk/u2.htm and at http://www.blackbirds
.net/u2/u2-timeline/u2tl160.html.
The black, single-engine craft bore little resemblance: A physical description
of the U-2 is available at http://www.airforce-technology.com/projects/u2/
and at http://www.area51specialprojects.com/genesis_u2.html.

116   filling the tanks from portable fifty-five-gallon oil drums: Richard Bissell, *Re-
flections of a Cold Warrior* (New Haven: Yale University Press, 1996), p. 118.
a 12,000-foot-long spool of high-resolution Kodak film: Michael Beschloss,
Mayday: Eisenhower, Khrushchev, and the U-2 Affair (New York: Harper and
Row, 1986), p. 92.
At 4:00 AM a doctor measured his temperature, pulse, and blood pressure:
Lieutenant Colonel Charles Wilson, USAF, "Flying the U-2," http://www
.blackbirds.net/u2/u-2mission.html.

117   Richard Bissell sat in his downtown office on H Street, across from the Metropolitan Club: Bissell, *Reflections of a Cold Warrior*, p. 98.

"We must find ways": James R. Killian, *Sputnik, Scientists, and Eisenhower: A Memoir of the First Special Assistant to the President for Science and Technology* (Cambridge, Mass.: MIT Press, 1977), p. 79.

Jewish grandparents had emigrated from the same part of Odessa: Philip Taubman, *Secret Empire: Eisenhower, the CIA, and the Hidden Story of America's Space Espionage* (New York: Simon and Schuster, 2003), p. 93.

118   Mark Twain House: Ibid., p. 115.

Allen Dulles was a significant shareholder: Grose, *Gentleman Spy*, pp. 371–73.

119   but Secretary Quarles had opted to go with a rival design by Bell Labs: Chris Pocock, "From the Shadows—Early History of the U-2," *Code One Airpower Magazine*, January 2002, http://www.codeonemagazine.com/archive/2002/articles/jan_02/shadows/index.html. See also Bissell, *Reflections of a Cold Warrior*, p. 93.

The contract was shrouded in such secrecy: The classified nature of the program is described by Garfield Thomas, vice president of ISR programs at Lockheed Martin Skunk Works, at http://www.area51specialprojects.com/genesis_u2.html.

Even within the White House staff, only two people: Dwight D. Eisenhower, *Waging Peace: The White House Years, 1956–61* (New York: Doubleday, 1965), p. 544.

120   "The ampoule should be crushed between the teeth": Taubman, *Secret Empire*, p. 178.

"We told Eisenhower": Bissell, *Reflections of a Cold Warrior*, p. 121.

Only three bolts . . . connected: Beschloss, *Mayday*, p. 92.

"Holy smokes": Taubman, *Secret Empire*, p. 131.

"Well, boys": Killian, *Sputnik, Scientists, and Eisenhower*, p. 84.

"If the Soviets ever capture one of these planes": Eisenhower, *Waging Peace*, p. 546.

121   Boyle's law: Wilson, "Flying the U-2."

"This is what would happen": March 9, 1955, "Tomorrowland" telecast, *Walt Disney Treasures*, Buena Vista Home Entertainment, Burbank, Calif., 2004, stock no. 31749.

122   "I was assured": Eisenhower, *Waging Peace*, p. 546.

123   "coffin corner": Wilson, "Flying the U-2."

Every U-2 flight required presidential approval: Beschloss, *Mayday*, p. 118.

A U-2 had exploded once: http://www.blackbirds.net/u2/u2-timeline/u2tl50.html.

124   "seemed somewhat startled and horrified to learn that the flight plan": Bissell, *Reflections of a Cold Warrior*, p. 112.

125   Only the new P-30 radar: Alexander Orlov, *The U-2 Program: A Russian Officer Remembers*, speech at 1998 CIA symposium, at http://www.cia.gov/csi/studies/winter98_99/art02.html.

"According to fully confirmed data": Sergei Khrushchev, *Nikita Khrushchev*, p. 159. Also author telephone interview with Sergei Khrushchev, February 10, 2006.

"What I remembered most": Author telephone interview with Sergei Khrushchev, February 10, 2006.

126   The forays, or "ferret" missions as they were known: Douglas Stanglin, Susan Headden, and Peter Cary, "Secrets of the Cold War: Special Report," *US News & World Report*, March 15, 1993, pp. 30–52.
       "Representations and recommendations": Ibid., p. 34.
       "It would have meant war": Sergei Khrushchev, *Nikita Khrushchev*, p. 157.

127   "The notion that we could overfly them": Bissell, *Reflections of a Cold Warrior*, p. 113.
       "For the first time": July 17, 1956, CIA memorandum, CIA-RDP62B00844R00200020017–3. Declassified August 26, 2000.

128   "A few days ago a super-long-range": Siddiqi, *Sputnik and the Soviet Space Challenge*, p. 161.

129   From its dish network atop a 6,800-foot peak: Burrows, *This New Ocean*, p. 225.

130   "Stop sending intruders into our air space": Orlov, *The U-2 Program*.
       "Father thirsted for revenge": Author telephone interview with Sergei Khrushchev, November 27, 2005.

131   "They blamed us": Richard Nixon, *RN: The Memoirs of Richard Nixon* (New York: Grosset and Dunlap, 1978), p. 182.
       a Ford repair shop at Fifth and K streets: Bissell, *Reflections of a Cold Warrior*, p. 104.
       "I don't 'believe' that the Soviets are ahead": *New York Times*, February 6, 1957.

132   "Every day we don't reverse our policy is a bad day for the Free World": Ibid., August 18, 1957.
       a West Coast military think tank, the RAND Corporation: Dickson, *Sputnik*, p. 46.
       "Gone was the folksy fellow": Burrows, *This New Ocean*, p. 145.

133   SR-71 Blackbird: http://www.sr-71.org/blackbird/sr-71/html.
       Bissell was alarmed that it was not even at the blueprint stage: Bissell, *Reflections of a Cold Warrior*, p. 134.

134   to personally inspect every Jupiter C launch: Bergaust, *Wernher von Braun*, p. 243.
       "I knew our national effort": Bissell, *Reflections of a Cold Warrior*, p. 134.

135   "in view of the competition we might face": http://www.history.nasa.gov/sputnik/chapter2.html.
       "It was unfortunate": Killian, *Sputnik, Scientists, and Eisenhower*, p. 14.
       "Ellender said that we must be out of our minds": Ward, *Dr. Space*, p. 96.
       "in the nearest future, the USSR will send a satellite into space": Mstislav Keldysh, ed., *Tvorcheskoye Naslediye Akademika Sergeya Pavlovicha Koroleva: Izbranye Trudy I Dokumenty* (Moscow: Nauka, 1980), p. 376.

136   "The creation and launching of the Soviet": F. J. Krieger, *Behind the Sputniks* (Washington, D.C.: Public Affairs Press, 1958), p. 329.
       "The Astronomical Council of the USSR": Ibid., p. 288.
       "It was hard not to feel that I was being set up": Nixon, *RN*, p. 167.
       "Dear Dick, I find that while I have thanked": Ibid., p. 181.

138   "He flashes gold cuff links": Randall B. Woods, *LBJ: Architect of American Ambition* (New York: Free Press, 2006), p. 338.
       "I'm going to have to bring up the nigger bill": Ibid., p. 326.
       "Southern whites are not bad people": James T. Patterson, *Brown v. Board of Education: A Civil Rights Milestone and Its Troubled Legacy* (New York: Oxford University Press, 2001), p. 81.

139   "What he had not done was provide leadership": Ambrose, *Eisenhower*, p. 409.
"I got the impression at the time": Halberstam, *The Fifties*, p. 686.

140   "We're trapped": And all other Little Rock quotes from inside high school quoted in Melba Patillo Beals, *Warriors Don't Cry* (New York: Pocket Books, 1994), pp. 87–88.
"The colored children [were] removed to their homes for safety purposes": The Mann telegram is available online at the Eisenhower Library under telegrammanntopresident92457 at http://www.eisenhower.archives.gov/dl/ LittleRock/littlerockdocuments.html.
"Troops not to enforce integration": Ibid., under DEtroopstoArkansas at http://www.eisenhower.archives.gov/dl/LittleRock/littlerockdocuments.html.
"A weak President": Halberstam, *The Fifties*, p. 987.

7: *A Simple Satellite*

142   The meeting had not gone as Korolev had hoped: The narrative of this account is drawn primarily from K. V. Gerchik, *Proryv v Kosmos* (Moscow: Veles, 1994), though the opinions of its participants are drawn from a wide array of sources listed below.

143   "I suggest we begin preparations to launch": Gerchik, *Proryv v Kosmos*, p. 29.
"This proposal was a big surprise": Ibid.
"All these space projects": Semenov, ed., *Raketno Kosmicheskaya Korporatsiya Energiya*, p. 87.

144   an unrealistic time frame: V. V. Favorskiy and I. V. Meshcheryakov, eds., *Voyenno-Kosmicheskiye Sily: Kosmonavtika I Vooruzhennyye Sily*, vol. 1 (Moscow: Sankt-Peterburgskoy Tipografia Nauka, 1997), p. 34.
"The Directorate of Missile Weapons": Yuri Mozzhorin, *Tak Eto Bilo* (Moscow: Tsnimash, 2000), p. 71.
"development of an artificial satellite for photographing the earth's surface": Vetrov, ed., *S. P. Korolev I Ego Delo*, p. 232.

145   who had been the youngest member ever elected to the prestigious Academy of Sciences: *Akademiya Nauk SSSR: Membership Directory*, vol. 2 (Moscow: Nauka, 1974), p. 61.
Korolev's greatest proponent was openly skeptical: Siddiqi, *Sputnik and the Soviet Space Challenge*, p. 154.

146   which, at Tyura-Tam, was just over 1,000 feet per second: Ivan V. Meshcheryakov, *V Mire Kosmonovtiki* (Novgorod: Russian Merchant Publishers, 1996), pp. 45–46.

147   "Why, Sergei Pavlovich?" and "Because it's not round": Golovanov, *Korolev*, p. 535.
The silver-zinc chargers alone weighed 122 pounds, providing power: Valentin Glushko, ed., *Kosmonavtika Entsiklopediya* (Moscow: Sovetskaya Entsiklopedia, 1985), pp. 290–91.
"Mindless malice": Cadbury, *Space Race*, p. 159.

148   "Do you know when Russia will build the bomb?": Halberstam, *The Fifties*, p. 25.
"German scientists in Russia did it": David McCullough, *Truman* (New York: Simon and Schuster, 1992), p. 748.

149   The mutineers had been dealt with: Taubman, *Khrushchev*, pp. 368–69.
"Nobody wanted to be accused of dragging their feet": Gerchik, *Proryv v Kosmos*, p. 30.

150   the commission formally informed the Kremlin that PS-1 was scheduled for liftoff on October 6, 1957: Siddiqi, *Sputnik and the Soviet Space Challenge*, p. 165.

It was model number 8k71PS, sixteen feet shorter: Timofei Varfolomeev, "Soviet Rocketry Conquered Space," part 1, *Spaceflight Magazine* (UK), August 1995, pp. 260–63.

151   "Silence fell whenever the Chief Designer appeared": Harford, *Korolev*, p. 129.

which sat on a felt-covered cradle in a sealed-off "clean room": Semenov, ed., *Raketno Kosmicheskaya Korporatsiya Energiya*, p. 90.

"Coats, gloves, it's a must": Cadbury, *Space Race*, p. 161.

Tikhonravov had pressurized the sphere with nitrogen: Glushko, ed., *Kosmonavtika Entsiklopediya*, pp. 290–91.

"I saw a crowd gathered around the satellite": Mozzhorin, ed., *Nachalo Kosmichiskoy Eri*, p. 23.

152   "What does it mean?": Galovanov, *Korolev*, pp. 537–38.

An overhead crane lifted the twenty-seven-ton empty shell: Semenov, ed., *Raketno Kosmicheskaya Korporatsiya Energiya*, p. 74.

153   "Well, shall we see off our first-born?": Golovanov, *Korolev*, p. 538.

A grainy and undated Soviet video: *Fifty Years of RKK Energya* (Moscow: RKK Energya [videotape], 1996).

over the next hour and ten minutes, the rocket was raised: Hubert Curien, *Baikonour* (Paris: Arnaud Colin Editeur, 1994), p. 147.

over the 120-foot-deep, five-football-fields-wide: Igor Barmin, *Na Zemle I V Kosmosy* (Moscow: VP Barmin Design Bureau of General Machine Building, 2001), p. 80.

Marshal Nedelin, in particular, was unhappy with the Soviet arrangement: Ibid., p. 93.

154   "Technical banditry": Golovanov, *Korolev*, p. 538.

"Let's not make a fuss": Ibid.

"OK, dear": A. Polyektov, *Kosmodrome Bajkonur; Nachale* (Moscow: Veles, 1992), p. 86.

155   "Nobody will rush us": Siddiqi, *Sputnik and the Soviet Space Challenge*, p. 165.

"T minus ten" and countdown instructions re-created from the following sources: Chekunov recollection in Gerchik, *Proryv v Kosmos*, pp. 68–73. Official timeline launch card reprinted in Natalia Koroleva, *Otets*, vol. 2, p. 309. Also from Chertok, *Raketi I Lyudi*, vol. 2, pp. 197–98. And from Ishlinskiy, ed., *Akademik S. P. Korolev*, pp. 448–64.

158   At 116 seconds a fiery cross appeared: *Novosti Kosmonavtiki*, no. 7, August 1997, p. 9.

The engines had run out of fuel at 295.4 seconds: Ibid.

at 142 miles in altitude instead of 147 miles: Ibid.

"Separation Achieved": Semenov, ed., *Raketno Kosmicheskaya Korporatsiya Energiya*, p. 89.

159   "Quiet": Golovanov, *Korolev*, p. 540.

"This is music no one has ever heard before": Cadbury, *Space Race*, p. 164.

"Hold off on the celebrations": Mozzhorin, ed., *Nachalo Kosmichiskoy Eri*, p. 64.

160   BEEP, BEEP, BEEP: Ibid.

8: By the Light of a Red Moon

161   a man so hated in Huntsville that some rocket scientists had once burned his effigy in Courthouse Square: Ward, *Dr. Space*, p. 98.
"We could not shed a single tear": Medaris, *Countdown for Decision*, p. 152.

162   "Sour Kraut Hill": Ward, *Dr. Space*, p. 78.
The jobs of five thousand skilled workers: Bergaust, *Wernher von Braun*, p. 218.
bureaucratic guerrilla campaigns that were beginning to take their toll: Killian, *Sputnik, Scientists, and Eisenhower*, p. 127.

163   "timing his comings and goings so that Grandmother": Harris, *A New Command*, p. 7.
over one hundred thousand dollars in his trading account: Ibid., p. 43.
"We must make it perfectly clear": Bille and Lishoke, *The First Space Race*, p. 117.
soaring 662 miles high over a 3,335-mile arc: McDougall, *The Heavens and the Earth*, p. 130.
"In various languages": Von Braun et al., *Space Travel*, p. 156.

164   At fifty-three, he was almost exactly: Neil McElroy's biographical information can be found at http://www.nndb.com/people/102/000057928.
"Our whole organization was thoroughly fired up": Medaris, *Countdown for Decision*, p. 154.
"two-star generals were serving drinks to three-star generals": Ward, *Dr. Space*, p. 111.

165   the young officer rudely interrupted McElroy: Medaris, *Countdown for Decision*, p. 155.
"General Gavin was visibly shaken": Ordway and Sharpe, *The Rocket Team*, p. 261.
"Damn bastards": Stuhlinger and Ordway, *Wernher von Braun*, p. 131.
"Now look": Ibid.

166   "Von Braun started to talk as if ": Medaris, *Countdown for Decision*, p. 155.
"We knew they would do it": Ibid.
"There was no chance": Bille and Lishoke, *The First Space Race*, p. 118.
"For God's sake, cut us loose": Medaris, *Countdown for Decision*, p. 155.

167   "It was imprudent to admit we had retained those rockets": Harris, *A New Command*, p. 155.
"It beeped derisively over our heads": Medaris, *Countdown for Decision*, p. 156.

168   "Missile number 27 proved our capabilities": Ibid.
"When you get back to Washington": McDougall, *The Heavens and the Earth*, p. 131.
64 percent, according to a Gallup survey: *Time*, October 14, 1957.

169   "Dear Dick, I had been hoping to play golf this afternoon": *The Papers of Dwight David Eisenhower*, vol. 18 (Baltimore: Johns Hopkins University Press, 1996), pt. 3, chap. 6, document 365.
"Sherman Adams was cold, blunt, abrasive": Nixon, *RN*, p. 198.
"Golf in Newport was enjoyable": *The Papers of Dwight David Eisenhower*, vol. 18, pt. 3, chap. 6, document 366.

170   "an economy of abundance": Halberstam, *The Fifties*, p. 587.
And tax revenues were coming in at a disappointing $72 billion: Eisenhower, *Waging Peace*, p. 213.
in an effort to trim half a billion dollars from the $3.5 billion monthly defense bill: Robert A. Divine, *The Sputnik Challenge: Eisenhower's Response to the Soviet Threat* (New York: Oxford University Press, 1993), p. 119.

170   "The developments of this year": Ferrell, ed., *Eisenhower Diaries*, p. 347.
the putting green he had installed just outside his patio doors: Ambrose, *Eisenhower*, p. 398.

171   played golf for the fifth time that week: Dickson, *Sputnik*, p. 22.
"an event of considerable technical and scientific importance": John Foster Dulles Papers, Dwight Eisenhower Library, Abilene, Kansas, at http://www.history.nasa.gov/sputnik/15.html.
"without military significance," "A neat technical trick," "A silly bauble," "in an outer space basketball game": Killian, *Sputnik, Scientists, and Eisenhower*, p. 10.
"hunk of iron": Burrows, *This New Ocean*, p. 187.
"a propaganda stunt": Divine, *The Sputnik Challenge*, p. xv.
like a "canary that jumps on the eagle's back": Burrows, *This New Ocean*, p. 186.
"Listen now": *Time*, October 12, 1956.
"Soviet Fires Earth Satellite": *New York Times*, October 5, 1957.

172   "Here in the capital": Richard Witkin, ed., *The Challenge of the Sputniks* (New York: Doubleday Headline Publications, 1958), p. 9.

173   "a great national emergency," comparisons to the shots fired at Lexington and Concord, "a technological Pearl Harbor": Divine, *The Sputnik Challenge*, p. xvi.
"chilling beeps": *Time*, October 14, 1957.
"The reaction here indicates massive indifference": Witkin, ed., *The Challenge of the Sputniks*, p. 3.
only 13 percent of Americans saw Sputnik as a sign: Dickson, *Sputnik*, p. 23.

174   "When I asked what this country should do": Shirley Ann Warshaw, ed., *Reexamining the Eisenhower Presidency* (Westport, Conn: Greenwood Press, 1993), p. 108.
"I have been warning about this growing danger": *New York Times*, October 5, 1957.
"a National Week of Shame and Danger": *Time*, October 21, 1957.
"We now know beyond a doubt": *New York Times*, October 5, 1957.

175   "liked nothing better than to career over the hills": Woods, *LBJ*, p. 313.
an air-conditioned, glass-enclosed, forty-foot-high hunting blind: Ibid.
"Soon they will be dropping bombs on us": Dickson, *Sputnik*, p. 117.
"a full and exhaustive inquiry": Henry Dethloff, *Suddenly, Tomorrow Came* (Washington: United States Government Printing Office, 1994), p. 3.

176   "I made sure that there was always one companion": Doris Kearns Goodwin, *Lyndon Johnson and the American Dream* (New York: St. Martin's Press, 1976), p. 105.
"I knew there was only one way to see Russell everyday": Ibid., p. 103.
some 4 percent of the U.S. population would report seeing Sputnik: International Affairs Seminars of Washington, "American Reactions to Crisis: Examples of Pre-Sputnik and Post-Sputnik Attitudes and of the Reaction to Other Events Perceived as Threats," U.S. President's Committee on Information Activities Abroad Records, 1959–1961, box 5, A83-10, Dwight D Eisenhower Library, Abilene, Kansas, at http://www.history.nasa.gov/sputnik/oct58.html.

177   "We can't always go changing our program": Warshaw, ed., *Reexamining the Eisenhower Presidency*, p. 109.
"Ike Plays Golf, Hears the News": Dickson, *Sputnik*, p. 22.
"This was a place where Eisenhower went wrong": Bille and Lishock, *The First Space Race*, p. 107.

177   "I can't understand why the American people": Ibid.

178   "There was no doubt that the Redstone" and other quotes from October 8, 1957, White House damage-control meeting: http://www.archives.gov/education/lessons/sputnik-memo/images/memo-page2-1.gif.

179   "The rocketry employed by our Naval Research Laboratory for launching our Vanguard" and all other quotes from October 9, 1957, press conference: Official White House transcript of President Eisenhower's Press and Radio Conference no. 123. October, 9, 1957, the Dwight D. Eisenhower Library, Abilene, Kansas, at http://www.eisenhower.archives.gov/dl/sputnik/pressconferenceoct91957pg.1pdf.

182   "A fumbling apologia": http://www.history.nasa.gov/sputnik/chap11.html.
       "A Crisis in Leadership": *Time*, November 4, 1957.
       "be in some kind of partial retirement": Dickson, *Sputnik*, p. 120.
       "He is not leading the country": Divine, *The Sputnik Challenge*, p. 8.
       mind-numbing sedatives: Dickson, *Sputnik*, p. 120.
       "courageous statesmanship": Ibid.
       "penny-pinching," "complacency," "lack of vision," "incredible stupidity": http://www.history.nasa.gov/sputnik/chap11.html.
       "No greater opportunity will ever be present": McDougall, *The Heavens and the Earth*, p. 149.
       "The issue of [Sputnik], if properly handled": Ibid.

183   "Its velocity was breathtaking": Woods, *LBJ*, p. 262.
       "making cowboy love": Ibid., p. 263.

184   Go to Congress, he urged Ike: McDougall, *The Heavens and the Earth*, p. 147.
       "We do not, as of yet, know if the satellite is sending out encoded messages" and all other quotes from October 10, 1957, NSC meeting: Memorandum on the 339th meeting of the National Security Council, Eisenhower Library, Abilene, Kansas, at http://www.Eisenhower.archives.gov/dl/sputnik/summaryofdiscussion339thmtgoct1119571of14.pdf.

186   "The country will support it": Divine, *The Sputnik Challenge*, p. 20.
       announcing that the Vanguard program would launch "a small satellite sphere": http://www.eisenhower.archives.gov/dl/sputnik/pressconferenceoct91957pg.1pdf.
       "We who could coldly appraise": Medaris, *Countdown for Decision*, p. 160.

187   "far, far out on a limb": Ibid., p. 162.
       "I had neither money nor authority": Ibid.

9: Something for the Holidays

188   "Just another Korolev launch": Nicholas Daniloff, *The Kremlin and the Kosmos* (New York: Alfred A. Knopf, 1972), p. 65; see also Harford, *Korolev*, p. 122.

189   The man whose popularity had so intimidated Joseph Stalin: Thumbnail bio of Zhukov is drawn from http://www.worldatwar.net/biography/z/Zhukov. See also Albert Axell, *Marshal Zhukov: The Man Who Beat Hitler* (London: Longman, 2003).

190   "Where you find Zhukov, you find Victory": http://www.worldatwar.net/biography/z/Zhukov.

191   Khrushchev had unilaterally slashed troop forces by a staggering 2 million men: Taubman, *Khrushchev*, p. 379.

191   "shark fodder": Ibid.
       a further round of three-hundred-thousand-troop reductions: Ibid.
       "Some voices of dissatisfaction were heard": Nikita Khrushchev, *Khrushchev Remembers*, vol. 2, p. 43.
       "I can't go to battle with generals who have to travel with field hospitals": Sergei Khrushchev, *Nikita Khrushchev*, p. 225.
192   "He assumed so much power": Nikita Khrushchev, *Khrushchev Remembers*, vol. 2, p. 44.
       "Father feared that Zhukov saw General Eisenhower as an example": Author telephone interview with Sergei Khrushchev, September 15, 2006.
       "I see what Zhukov is up to": Nikita Khrushchev, *Khrushchev Remembers*, vol. 2, p. 44.
       "saboteur schools": Author telephone interview with Sergei Khrushchev, September 15, 2006.
       "a South American–style military takeover": Nikita Khrushchev, *Khrushchev Remembers*, vol. 2, p. 44.
193   "His unreasonable activities leave us no choice": Ibid., p. 45.
194   a small and unobtrusive squib would appear on the back page of *Pravda*: Sergei Khrushchev, *Nikita Khrushchev*, p. 250.
       The evening sessions had run well past ten: The account of the meeting in Kiev is drawn from telephone interviews with Sergei Khrushchev in September 2006 and from his memoir.
195   "I'll be back": Sergei Khrushchev, *Nikita Khrushchev*, pp. 259–60; also author telephone interview with Sergei Khrushchev, September 15, 2006.
196   the lead story in *Pravda* on the morning of October 5: Golovanov, *Korolev*, p. 533.
197   orbital decay: I. V. Meshcheriakov, *V Mire Kosmonavtiki* (Nizhny Novgorod: Russki Kupets, 1996), pp. 35–36.
       crossing the equator every ninety-six minutes at a sixty-five-degree angle: Valentin Glushko, ed., *Malenkaya Entsiklopedia Kosmonavtiki* (Moscow: Sovetskaya Entsiklopedia, 1970), p. 520.
       The exact number would turn out to be ninety-two days: Ibid.
198   "We were all too focused": Golovanov, *Korolev*, p. 543.
       "It was late": Chertok, *Rakety I Lyudi*, vol. 2, p. 195.
       "We thought the satellite": Bille and Lishock, *The First Space Race*, p. 104.
       "This date": Burrows, *This New Ocean*, p. 197.
       "The whole world is abuzz": Golovanov, *Korolev*, p. 524.
199   green for foreign press clippings, red for decoded diplomatic traffic: Sergei Khrushchev, *Khrushchev on Khrushchev*, p. 128.
       "The achievement is immense": *Manchester Guardian*, October 6, 1957.
       "Myth has become reality": *Le Figaro*, October 7, 1957.
       "A turning point in civilization": *New York Times*, October 6, 1957.
       "in contrast with the first steps in the atomic age": Monographs in Aerospace History no. 10: USIA, October 17, 1957, Report on World Opinion, at http://www.history.nasa.gov/45thann/html/pubs.
       "validation of the superiority of Marxist-Leninist technology": Ibid.
       "the planetary era rings the death knell of colonialism": Dickson, *Sputnik*, p. 131.
200   "With only a ball of metal": Siddiqi, *Sputnik and the Soviet Space Challenge*, p. 171.
       The European Assembly in Strasbourg . . . and other examples of shaken faith in the United States: Monographs in Aerospace History no. 10: USIA,

October 17, 1957, Report on World Opinion, at http://www.history.nasa
.gov/45thann/html/pubs.

200   "Public opinion in friendly countries shows decided concern": Ibid.

201   "People all over the world are pointing to the satellite": *Time*, October 21,
1957.

"World's First Artificial Satellite of the Earth Created in Soviet Union":
Pravda, October 6, 1957.

202   "The average American only cares for his car": Harris, *A New Command*, p. 182.

"We could never understand": David Akens, *Army Ballistic Missile Agency Historical Monograph* (Huntsville, Ala.: Redstone Arsenal, 1958), appendix A, at
http://www.www.army.redstone.mil/history.

203   he astounded Korolev by asking where the satellites were placed: Siddiqi,
Sputnik and the Soviet Space Challenge, p. 169.

"People in the Soviet Union did not complain during that era": Author telephone interview with Sergei Khrushchev, September 15, 2006.

"I remember walking in Red Square": Natalia Koroleva, interview in televised
documentary film *The Secret Designer* (Toronto: Ryan Productions, 1994).

"They are well provided for": Daniloff, *The Kremlin and the Kosmos*, p. 128.

"Our most brilliant missile designer could not hold a candle to Sergei
Pavlovich Korolev": Nikita Khrushchev, *Khrushchev Remembers*, vol. 2, p. 77.

204   "When we announced the successful testing of an intercontinental rocket":
McDougall, *The Heavens and the Earth*, p. 237.

205   "Initially, Father believed": Author telephone interview with Sergei
Khrushchev, September 15, 2006.

Yangel, a few months earlier, had successfully tested the R-12: *Oruzhe Rossii*,
vol. 4 (Moscow: Military Parade, 1997), p. 77.

lobbying Nedelin to push for the R-16: Cadbury, *Space Race*, p. 182.

206   "In his able proposals": Daniloff, *The Kremlin and the Kosmos*, p. 72.

"You know": Khrushchev's exchange with Korolev and Mikoyan is in Golovanov, *Korolev*, p. 544.

207   Khrushchev had commissioned poems: Harford, *Korolev*, p. 122.

commemorative stamps: Golovanov, *Korolev*, p. 542.

"Now we are ahead of America": Witkin, ed., *The Challenge of the Sputniks*, p. 71.

208   "Nowhere else would you find": Ibid.

Beijing had blasted Khrushchev's assault on Stalin as "revisionist": Medvedev
and Medvedev, *Khrushchev*, p. 72.

promised Mao missile technology, starting with the R-2: Semenov, ed.,
Raketno Kosmicheskaya Korporatsiya Energiya, p. 66.

209   He had the parts to assemble one more rocket: Chertok, *Rakety I Lyudi*, vol. 2,
p. 199.

211   The hardware would have to come entirely off the shelf: Harford, *Korolev*,
p. 132.

212   "My wife and I": in Mozhorin, ed., *Nachalo Kosmicheskoy Eri*, p. 64.

"We're returning to Tyura-Tam": Ibid.

10: Operation Confidence

213   "What next?": Divine, *The Sputnik Challenge*, p. 94.

"Shoot the Moon, Ike": *Time*, November 11, 1957.

213   "Plunge heavily into this one": http://www.spacereview.com/article/396/1.
214   "Let's not look for scapegoats": *Legislative Origins of the National Aeronautics and Space Act: Proceedings of an Oral History Workshop Conducted April 3, 1992*, Monographs in Space History no. 8, http://www.history.nasa.gov/40than/legislat.pdf.
215   "Sputnik II absolutely made the decision for them": Ibid.
       "The greatly increased size of the second Sputnik": *Time*, November 11, 1957.
       "whether the Soviet Union might be using some new form": Dickson, *Sputnik*, p. 143.
       "As Chairman of the Committee": www.spacereview.com/article/396/1.
       "It's a real circus act": Divine, *The Sputnik Challenge*, p. 44.
216   "demonstrates that the USSR has outstripped": Daniloff, *The Kremlin and the Kosmos*, p. 128.
       "the freed and conscientious labor of the people": Bille and Lishock, *The First Space Race*, p. 103.
       "to be less concerned with the depth of the pile": Dickson, *Sputnik*, p. 140.
       "While we devote our industrial and technological might": Witkin, ed., *The Challenge of the Sputniks*, p. 19.
       "It's time to stop worrying about tail-fins": Witkin, ed., *The Challenge of the Sputniks*, p. 77.
       "We've become a little too self-satisfied": Dickson, *Sputnik*, p. 139.
       "an intercontinental outer-space raspberry": Witkin, ed., *The Challenge of the Sputniks*, p. 17.
217   "From the echoes of the satellite": Warshaw, ed., *Reexamining the Eisenhower Presidency*, p. 111.
       "The fact that we were able to launch the first Sputnik": Daniloff, *The Kremlin and the Kosmos*, p. 127.
       "The United States can practically annihilate": Eisenhower, *Waging Peace*, p. 223.
218   GENTLE IN MANNER, STRONG IN DEED: Killian, *Sputnik, Scientists, and Eisenhower*, p. 219.
       "It misses the whole point": Warshaw, ed., *Reexamining the Eisenhower Presidency*, p. 112.
       General Bruce Medaris watched the address with an equal mix of bewilderment and frustration: Medaris, *Countdown for Decision*, p. 165.
219   "somewhat cherubic" and "as disarmingly pleasant": Taubman, *Secret Empire*, p. 88.
       Squirrel Hill: Harris, *A New Command*, p. 147.
       "Hang on tight, and I will support you": Medaris, *Countdown for Decision*, p. 165.
       seriously considering quitting the army: Ibid., p. 168.
220   "So far as the public could judge": Ibid., p. 166.
       "The time for talking": Ibid., p. 169.
       "The real tragedy of Sputnik's victory": Stuhlinger and Ordway, *Wernher von Braun*, p. 132.
       "could be very damaging to what the President was trying to do": Medaris, *Countdown for Decision*, p. 169.
       a devastating report: Prados, *The Soviet Estimate*, p. 72.
221   "deeply shocking": Sherman Adams, *First Hand Report: The Story of the Eisenhower Administration* (New York: Harper and Brothers, 1961), p. 413.

221   "Its disclosure would be inimical": Eisenhower, *Waging Peace*, p. 221.
"It will be interesting to find out how long": Ibid.
"The still top-secret Gaither Report": Killian, *Sputnik, Scientists, and Eisenhower*, p. 98.

222   "Arguing the Case for Being Panicky": McDougall, *The Heavens and the Earth*, p. 150.
"Another tranquility pill": Divine, *The Sputnik Challenge*, p. 47.
"It was by no means a blood, sweat and toil speech": Witkin, ed., *The Challenge of the Sputniks*, p. 34.
"Two Sputniks cannot sway Eisenhower": Ibid, pp. 45–46.
sinking by 22 percentage points: Dickson, *Sputnik*, p. 151.
"In a matter of a few months": McDougall, *The Heavens and the Earth*, p. 156.
"The bill's best bet": Ibid., p. 161.
"Eisenhower was skeptical about the loans": Killian, *Sputnik, Scientists, and Eisenhower*, p. 195.

223   A new $100-million-a-year Astronautical Research and Development Agency: Bille and Lishock, *The First Space Race*, p. 112.
"I'd like to know what's on the other side of the moon": Ambrose, *Eisenhower*, p. 453.
"a depression that will curl your hair": Greenstein, *The Hidden-Hand Presidency*, p. 121.
Unemployment was expected to jump by as much as 1.5 million: Eisenhower, *Waging Peace*, p. 213.

224   "In effect there was no clear cut authority": Medaris, *Countdown for Decision*, p. 167.
"They are trying to delude Congress": Harris, *A New Command*, p. 183.
"Either give me a clear-cut order": Stuhlinger and Ordway, *Wernher von Braun*, p. 134.
"I'm afraid my language": Medaris, *Countdown for Decision*, p. 168.
"a fierce religious zeal" and a "pious belligerence": Killian, *Sputnik, Scientists, and Eisenhower*, p. 127.

225   "Vanguard will never make it": Medaris, *Countdown for Decision*, p. 155.
"all test firings of Vanguard have met with success": Ibid., p. 166.
stop sending him "garbage": Kurt Stehling, *Project Vanguard* (New York: Doubleday, 1961), p. 119.
"almost developed": Ibid., p. 60.
"For all practical purposes the Vanguard vehicle was new": Constance McLaughlin Green and Milton Lomask, *Vanguard: A History* (Washington, D.C.: NASA, 1970), p. 177.

226   "It was either forgotten, or not understood": Stehling, *Project Vanguard*, p. 60.
simultaneously drew paychecks from the aerospace companies: Bergaust, *Wernher von Braun*, p. 240.
assertions from the Glenn L. Martin Company: Green and Lomask, *Vanguard*, pp. 54, 62.
Vanguard's budget: Ibid., pp. 62, 105, 131.
"I question very much whether it would have been authorized": Percival Brundage, April 30, 1957, Project Vanguard memorandum to the president, Bureau of Budget files, Dwight D. Eisenhower Library, Abilene, Kansas, at http://www.history.nasa.gov/sputnik/iik4.html.

227   "piece by rotten piece": Stehling, *Project Vanguard*, p. 119.

227   There were moisture problems, poorly located pressure indicator lines, unsoldered wire connections, corroded and leaky fittings: Ibid., pp.109–11.
      "What! You want to put a ball in that rocket?": Ibid., pp. 87–88.
      "We're never going to make it": Green and Lomask, *Vanguard*, p. 131.
228   "an unaccepted, incompletely developed vehicle": Ibid., p. 177.
      "An astonishing piece of stupidity": *Time*, October 21, 1957.
      the Stewart Committee had been "prejudiced": Stehling, *Project Vanguard*, p. 60.
229   "the funds estimated by Secretary Quarles were totally inadequate": Witkin, ed., *The Challenge of the Sputniks*, p. 21.
      Wilson interviewed by Mike Wallace: Ibid., p. 47.
      "Implicit in all the criticism": Ambrose, *Eisenhower*, p. 457.
      a crack team of Wall Street lawyers: Robert A. Caro, *Master of the Senate: The Years of Lyndon Johnson* (New York: Alfred A. Knopf, 2002), p. 1022.
230   "He never asked the head of my organization": Eilene Galloway, NASA Oral History transcript, at http://www.jsc.nasa.gov/history/oral_histories/NASA_HQ/Herstory/GallowayE/EG_8-7-00.pdf.
      "He was really like a dynamo": Ibid.
      "The timing was perfect": *Legislative Origins of the National Aeronautics and Space Act: Proceedings of an Oral History Workshop Conducted April 3, 1992*, Monographs in Space History no. 8, http://www.history.nasa.gov/40than/legislat.pdf.
      "Crisis had become normalcy": Eisenhower, *Waging Peace*, p. 226.
      "His aides who sometimes caught him with a faraway look": Mosley, *Dulles*, p. 439.
      Gallup polls had shown that most American voters did not mind Ike's frequent weekday golf outings: Greenstein, *The Hidden-Hand Presidency*, p. 40.
231   "Oh little Sputnik, flying high": Roger D. Launius, *Sputnik and the Origins of the Space Age*, monograph at http://www.history.nasa.gov/sputnik/sputorig.html.
      "As I picked up a pen": Eisenhower, *Waging Peace*, p. 227.
      "You may be President in twenty-four hours": Nixon, *RN*, p. 184.
232   "The Vanguard tower was clear against a starry sky": Green and Lomask, *Vanguard*, p. 206.
      "Bird Watch Hill": Bille and Lishock, *The First Space Race*, p. 122.
233   filled the airwaves with all manner of facts: Ibid.
      Though missiles had been tested at the complex since the summer of 1950: http://www.patrick.af.mil/library/factsheets/factsheet_print.asp?fsID=4514&page=1.
      "The rocket looked unkempt": Stehling, *Project Vanguard*, p. 21.
234   the most ambitious and expensive installment of his "Man in Space" series: Introduction by Leonard Maltin to *Tomorrowland: Disney in Space and Beyond*, commemorative DVD package, Buena Vista Home Entertainment, Burbank, Calif., 2004, originally aired December 4, 1957.
235   Wolfsschanze: This account of the July 1943 meeting is drawn from Neufeld, *The Rocket and the Reich*, p. 192.
      and film that had been shot using several cameras simultaneously: Ward, *Dr. Space*, p. 33.
236   "seemed a pretty dowdy type": Piszkiewicz, *Wernher von Braun*, p. 27.
      somewhat more reluctant decision in 1940: Stuhlinger and Ordway, *Wernher von Braun*, p. 41.

237   "But what I want is annihilation": Neufeld, *The Rocket and the Reich*, p. 192.
"The Führer was amazed at von Braun's youth": Ibid., p. 278.
238   Porter was instrumental in scuttling: Bergaust, *Wernher von Braun*, p. 240.
239   "Ten, nine, eight . . .": Green and Lomask, *Vanguard*, pp. 208–9.
"Oh God! No! Look out! Duck!": Stehling, *Project Vanguard*, p. 24.

11: Goldstone Has the Bird

240   "What happened yesterday has made us . . ." and all other quotes from NSC meeting no. 347: http://www.eisenhower.archives.gov/dl/Sputnik/nasasearchreport.pdf, see NSC series, box 9, Ann Whitman File.
where polls conducted in Britain and France prior to Sputnik's launch had shown that only 6 percent: McDougall, *The Heavens and the Earth*, p. 240.
241   "Why doesn't somebody go out there, find it, and kill it": Piszkiewicz, *Wernher von Braun*, p. 117.
"Oh, what a Flopnik": Dickson, *Sputnik*, p. 158.
Kaputnik, Splatnik, Stallnik: *Time*, December 16, 1957.
"This incident has no bearing on our programs": Divine, *The Sputnik Challenge*, p. 72.
242   Sputnik Cocktails: Witkin, ed., *The Challenge of the Sputniks*, p. 4.
"our worst humiliation since Custer's last stand": *Time*, December 16, 1957.
Already, the Glenn L. Martin Company: Green and Lomask, *Vanguard*, p. 210.
"This program has had unprecedented publicity": *Time*, December 16, 1957.
"It lies with the President": Piszkiewicz, *Wernher von Braun*, p. 117.
"How long, how long oh God": Cadbury, *Space Race*, p. 173.
95 percent of his speech and motor skills: Eisenhower, *Waging Peace*, p. 229.
"There were open and widespread suggestions that the President resign": *Time*, December 9, 1957.
"It is the whole free world that is sick in bed with Ike": Divine, *The Sputnik Challenge*, p. 59.
243   "In my mind was the question": Eisenhower, *Waging Peace*, pp. 230–31.
The move would inspire a running joke: Dickson, *Sputnik*, p. 144.
"This man is not what he was": Killian, *Sputnik, Scientists, and Eisenhower*, p. 234.
245   The Chief Designer was diagnosed with arrhythmia, coupled with "over-fatigue": Cadbury, *Space Race*, p. 175.
a region-wide search for Tsander's grave: Ibid.
246   tales of the mixed-breed terrier Laika: Harford, *Korolev*, p. 132.
astonishingly detailed descriptions of the devices: Burrows, *This New Ocean*, p. 207.
"Tell me, Sergei Pavlovich": Nikita Khrushchev, *Khrushchev Remembers*, vol. 2, p. 82.
247   "to deliver the Soviet Coat of Arms to the Moon": Cadbury, *Space Race*, p. 179.
"Korolev works for TASS": Harford, *Korolev*, p. 116.
"Before you begin your questioning": Medaris, *Countdown for Decision*, p. 178.
248   "The chief reason": Philip Nash, *The Other Missiles of October: Eisenhower, Kennedy, and the Jupiters, 1957–1963* (Chapel Hill: University of North Carolina Press, 1997), p. 20.

248   "was being authorized to proceed on a 'top-priority' basis": Caro, *Master of the Senate*, p. 1025.
       "With feelings much different from those": Medaris, *Countdown for Decision*, p. 190.
249   "Proposal for a National Integrated Missile and Space Vehicle Development Program": Ordway and Sharpe, *The Rocket Team*, p. 262.
       "like a yo-yo": NASA Oral History series, http://www.jsc.nasa.gov/history/oral_histories/NASA_HQ/Ballistic/SchrieverBA/schrieverba.pdf.
       "Sputnik woke us up": Ibid.
       "So I also closed the door": Medaris, *Countdown for Decision*, p. 188.
       tried to covertly buy the Itek Corporation: Taubman, *Secret Empire*, p. 229.
       In the summer of 1957, Bissell, Edwin Land, and Jim Killian: Ibid., p. 230.
250   "Our first goal was to put the genie back in the bottle": Bissell, *Reflections of a Cold Warrior*, p. 135.
       "I had to invent an elaborate cover explanation": Ibid., p. 136.
251   "Some awful needles were stuck into this thing": *Legislative Origins of the National Aeronautics and Space Act: Proceedings of an Oral History Workshop Conducted April 3, 1992*, Monographs in Space History no. 8, http://www.history.nasa.gov/40than/legislat.pdf.
       "We will be walking a very tight wire with our lives": Bergaust, *Wernher von Braun*, p. 226.
252   "There are too many people in government who have the right to say no": Harris, *A New Command*, p. 191.
       "At the Pentagon they shudder when they speak of the 'gap' ": Prados, *The Soviet Estimate*, p. 80.
       "a leaky ship, with a committee on the bridge": Divine, *The Sputnik Challenge*, p. 118.
       "Speaking so fast": Caro, *Master of the Senate*, p. 1023.
       "Control of space means control of the world": Goodwin, *Lyndon Johnson and the American Dream*, p. 145.
253   "Light a match behind Lyndon and he'll orbit": Caro, *Master of the Senate*, p. 1025.
       "I'm not going to ask you about the precise date": Green and Lomask, *Vanguard*, p. 214.
       "Do not admit to the presence of the vehicle": Harris, *A New Command*, p. 184.
       "I desire it well understood": Bille and Lishock, *The First Space Race*, p. 127.
       "Personal observation had convinced me": Medaris, *Countdown for Decision*, p. 193.
254   It consisted of four stages: http://www.history.msfc.nasa.gov/milestones/chpt4.pdf.
       extending the Redstone's burning time from 121 to 155 seconds: http://www.nasa.gov/mission_pages/explorer/index.html.
       "Ship it to Florida, it will do the job": Bergaust, *Wernher von Braun*, p. 275.
255   "It became quite obvious that every effort": Medaris, *Countdown for Decision*, p. 196.
       "We bootlegged the whole job": Dickson, *Sputnik*, p. 169.
       "thought it would be wise to prepare it in such a way": Stuhlinger and Ordway, *Wernher von Braun*, pp. 134–35.
256   "Almost every reference to Army-developed hardware was stricken": Medaris, *Countdown for Decision*, p. 196.

256   "This is our biggest challenge": Ibid., p. 188.
257   "just about as thoroughly bored": Nash, *The Other Missiles of October*, p. 35.
       "The symbols of 1957 were two pale, clear streaks of light": *Time*, January 6, 1958.
       "make some specific arrangements": Eisenhower, *Waging Peace*, p. 223.
       "I decided to confine the annual message": Ibid., p. 240.
258   "We could see the Army preparations on their launch pad": Stehling, *Project Vanguard*, p. 159.
259   "The night was miserable cold and wet": Ibid., p. 163.
       "Our people did not take kindly to the idea of sitting": Medaris, *Countdown for Decision*, p. 200.
260   "Above our meeting in the hangar hovered a ghostly": Stehling, *Project Vanguard*, p. 175.
261   "an ardent Nazi" who had "denounced his colleagues to the Gestapo": Cadbury, *Space Race*, p. 12.
       polysulfide aluminum and ammonium perchlorate: http://www.history.nasa .gov/sputnik/expinfo.html.
262   apex predictor: Stuhlinger and Ordway, *Wernher von Braun*, p. 136.
       "Do you really want to rely on this alone?": Ibid., p. 137.
       winds . . . reaching 225 miles per hour: Ordway and Sharpe, *The Rocket Team*, p. 263.
263   "What's happened? What are you going to do?": Medaris, *Countdown for Decision*, p. 207.
264   "Highly marginal. We do not recommend that you try it": Ibid., p. 209.
       winds . . . still gusted at 157 miles per hour: Bergaust, *Wernher von Braun*, p. 276.
       "Everyone was going on sheer nerve": Medaris, *Countdown for Decision*, p. 210.
       a twenty-four-year-old first lieutenant by the name of John Meisenheimer: Harris, *A New Command*, p. 187.
       "Every man on the crew was conscious that the hopes of a Nation": Medaris, *Countdown for Decision*, p. 212.
265   "The searchlights are going on and lighting up the vehicle": *Time*, February 10, 1958.
       "There is nothing that I have ever encountered": Medaris, *Countdown for Decision*, p. 212.
       "When the countdown reaches zero, the bird will not begin to rise immediately": *Time*, February 10, 1958.
266   "Go, baby! Go!": Harris, *A New Command*, p. 189.
       "No. Let 'em sweat a little": *Time*, February 10, 1958.
267   "I'm out of coffee and running low on cigarettes": Harris, *A New Command*, p. 189.
       "Do you hear her?". . . Do you hear her now?": Bergaust, *Wernher von Braun*, p. 278.
       "Wernher, what's happened?": Ibid.
       "*Goldstone has the bird!*": Medaris, *Countdown for Decision*, p. 224.

Epilogue

269   "It represented only a symbolic counter threat to the United States": Nikita Khrushchev, *Khrushchev Remembers*, vol. 2, p. 80.

"It would have been better to dump them in the sea": Nash, *The Other Missiles of October*, p. 3.

270   "It is entirely possible that having a failure in the oxygen equipment": http://www.eisenhower.archives.gov/dl/u2incident/departmentstatementon U25560.pdf.

at 8,500 feet by grappling hooks attached to the front of a C-119 military plane: Taubman, *Secret Empire*, p. 321.

271   "Those friggin missiles": Nash, *The Other Missiles of October*, p. 3.

a catastrophic explosion: Mikhail E. Kuznetsky, *Bajkonur, Korolev, Yangel* (Voronezh: Voronezh, 1997), p. 127.

272   So poor were the harvests that the Soviet Union faced food shortages: Medvedev and Medvedev, *Khrushchev*, pp. 118–19.

197 of the 200 full members: Ibid., p. 172.

273   Soviet industrial growth began to slow dramatically in the mid-1960s: Haines and Leggett, eds., *CIA's Analysis of the Soviet Union, 1947–1990*, p. 191.

"I did everything I could to patch up their friendship": Nikita Khrushchev, *Khrushchev Remembers*, vol. 2, p. 79.

the cause of death was complications during routine surgery: Harford, *Korolev*, p. 279.

274   more commonly known as the Scud: *Russian Arms Catalog*, vol. 4, *Strategic Missile Forces* (Moscow: Military Parade Publishers, 1997), pp. 44–45 (in English).

because he refused to endorse escalating America's involvement in Vietnam: http://www.gwu.edu/~erpapers/mep/displaydoc.cfm?docid=erpn-stusym.

275   "You want to threaten? We will answer threats": Dunar, *America in the Fifties*, p. 294.

The *hidden hand*: Greenstein, *The Hidden-Hand Presidency*, p. vii.

a reevaluation that coincided, in part, with the declassification: Dunar, *America in the Fifties*, p. 326.

276   His farewell address to the nation in January 1961: Eisenhower, *Waging Peace*, p. 615.

Donald Quarles . . . was to have succeeded Neil McElroy as secretary of defense in 1959: http://www.arlingtoncemetery.net/donaldau.htm.

Joseph P. Kennedy offered Medaris a job: Harris, *A New Command*, p. 213.

he accepted the presidency of the Lionel Corporation: Ibid., p. 216.

he was ordained an Episcopal priest: Ibid., p. 254.

Paris Match, the glossy French magazine, published a glowing article: Piszkiewicz, *Wernher von Braun*, pp. 163–64.

Arthur Rudolf . . . was quietly extradited to West Germany: Simpson, *Blowback*, p. 39.

278   it supplies the boosters that orbit private U.S. satellites: Author telephone interview with Paula Korn, Sea Launch Public Affairs Office, November 11, 2005.

ACKNOWLEDGMENTS

Though my name is on the cover, this book is really the product of two brilliant minds. It was conceived by Scott Waxman and Paul Golob, who cajoled, conspired, and finally convinced me that I was the right person to write it. And for this I am eternally grateful—both for their collective wisdom and for their confidence.

Scott is the personification of the proactive agent, and I'm truly lucky that he represents me. Paul is the best editor I've ever worked with: cerebral, classy, and a man of his word. His guiding hand can be felt throughout the manuscript, and his uncanny ability to see the big picture has endowed this project with whatever weight it has. As with Scott, I count myself fortunate to have worked with him.

At Times Books and Henry Holt, a number of other people have made this book possible: first and foremost, John Sterling, whose patronage has been instrumental. David Wallace-Wells, Chris O'Connell, Jessica Firger, Claire McKinney, Maggie Richards, and Nick Caruso also richly deserve praise. They are responsible for the finished product and its placement in bookstores. Farley Chase of the Waxman Agency has also been responsible for shaping this book, and he has worked tirelessly developing its audio and foreign editions. I owe him thanks, in several languages.

In the course of my research, I have greatly benefited from the accumulated knowledge of Peter Gorin, who is probably the leading civilian expert in the United States on Soviet space and missile programs. Peter's encyclopedic memory and vast archive of Russian scientific materials are the driving force behind the technical sections of the book. I would also be remiss if I did not thank Sergei Khrushchev, whose memory and recollections form the foundation for some the book's best action scenes. Professor Khrushchev—himself a rocket scientist, and one of the only surviving eyewitnesses to many of the events depicted in the text—was exceedingly generous with his time, and I owe him a great debt for sharing his political insights as well.

On the political front, Eric Rubin—a friend, senior foreign service officer, old Russia hand, and student of history—nudged me whenever I strayed in trying to make sense of American domestic and cold war politics in the 1950s, and I am grateful for his counsel. In re-creating the political dynamics of the era, I've also relied on the accrued wisdom of many American scholars and space historians. They are too numerous to cite here, but I've made a point of acknowledging their contributions whenever possible within the body of the text.

And last, but certainly not least, I want to thank Roberta, my muse, critic, fan, editor, and better half. She makes everything possible.

INDEX

About the Author

MATTHEW BRZEZINSKI is a former Moscow correspondent for *The Wall Street Journal* and has reported extensively on homeland-security issues for *The New York Times Magazine* and other publications. He is the author of *Casino Moscow: A Tale of Greed and Adventure on Capitalism's Wildest Frontier* and *Fortress America: On the Front Lines of Homeland Security*. He lives in Washington, D.C.